THE Health & Safety Guide FOR Film, TV & Theater

THE Health & Safety Guide FOR Film, TV & Theater

MONONA ROSSOL

ALLWORTH PRESS
NEW YORK

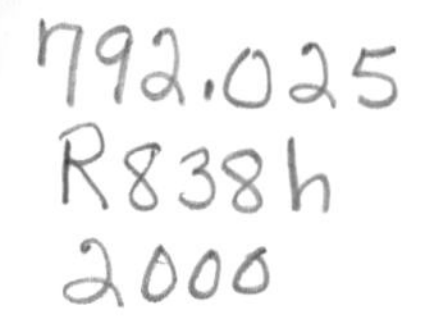

05 04 03 02 01 00 5 4 3 2 1

Published by Allworth Press
An imprint of Allworth Communications
10 East 23rd Street, New York, NY 10010

Cover design by Douglas Design Associates, New York, NY

Page composition/typography by SR Desktop Services, Ridge, NY

ISBN: 1-58115-071-7

Library of Congress Cataloging-in-Publication Data:

Rossol, Monona.
The health and safety guide for film, TV , and theater / Monona Rossol.
p. cm.
Includes bibliographical references and index.
ISBN 1-58115-071-7
1. Theaters—Health aspects. 2. Theaters—Safety measures. 3. Television broadcasting—Health aspects. 4. Performing arts—Health aspects. I. Title.

RC965.T54 R668 2001
363.11'9792—dc21

00-053591

Printed in Canada

Table of Contents

Read Me: An Introduction

Theatrical, motion picture, television, and entertainment productions are hazardous endeavors. In fact, it would be hard to imagine an industry with more hazards.

HAZARDS

During production, workers face the same hazards that other construction workers face: Riggers and lighting people climb about in the fly or work high atop powered lifts; woodworkers and welders build sets; and painters brush and spray color. Less obvious hazards are occurring in other shops where costumers use toxic dyes and solvents and where wigmakers and makeup artists work with the same cosmetics and sprays that cause high rates of illness in commercial beauticians and hairdressers.

On stage, actors emote their hearts out inches away from the ten-foot drop into the orchestra pit; fire, smoke, and explosions occur; split-second scene changes take place in the dark; and Peter Pan soars overhead on a thin line.

Many theater and entertainment professionals take these risks too much for granted. They would prefer no interference from safety experts or governmental officials. Theater safety should be, in their opinion, a subject discussed discreetly only among themselves.

Unfortunately, their preferences become moot points when safety is openly debated (or litigated) after accidents are reported in the press. But then public attention wanes, the accidents and illnesses are forgotten, and their causes remain essentially unchanged.

HEALTH V. SAFETY HAZARDS

Safety hazards are easier to explain than health hazards. Falls, cuts, and other accidents clearly demonstrate that *safety* hazards exist. *Health* hazards are less obvious and harder to prove.

Health hazards in theater, like industrial health hazards, result when individuals are exposed to toxic chemicals, fumes, and dusts or to physical phenomena such as ultraviolet or infrared light, noise, vibration, and excessive heat and cold.

Massive exposure to one of these health hazards may result in a serious illness or even death. However, it is more likely that repeated small exposures over weeks, months, or years will affect individuals gradually. Often, by the time an illness manifests itself, the damage is serious, sometimes irreversible.

But regardless of whether the risks are to our safety or long- or short-term health, there already are standards designed to protect us.

HEALTH AND SAFETY STANDARDS

The basic structure within which our health and safety problems are best addressed already exists. In fact, we are required to address them in this fashion by law. This structure is provided by the Occupational Safety and Health Administration (OSHA) regulations.

This book, unlike some theatrical magazine articles and books I've seen, will not make recommendations without making every effort to ensure that they are consistent with the laws. For this reason, OSHA regulations will be cited in the text from the Code of Federal Regulations (CFR) in their typical format (e.g., 29 CFR 1910.503(1)(a)).

I have included the OSHA citations to aid readers who need to look up the full regulation. I also hope the citations will make it clear that most of the book's recommendations are not personal ideas of mine. And since most of the book's recommendations are already required by law, readers should consider that ignoring them can result in citations, fines, and/or increased liability.

Some readers of this book may think that the OSHA regulations are irrelevant since they are not subject to them. The only people exempt from all forms of OSHA regulations are those who are self-employed and have no employees of their own. Even then, the exemption applies only when they are working in their own private studios. If they leave the studio to work as independent contractors in theaters or on locations, then they, too, must conform to the OSHA regulations at that work site. And as independent contractors, they may be held liable for accidents caused by their failure to conform to the regulations.

WHY THE OSHA REGULATIONS?

Every aspect of workplace safety and health is regulated by OSHA. (See also chapter 2 to determine if you come under a federal or state OSHA.) And readers should also know that in over twenty years of inspections and worker trainings that I have done in theaters and on film and TV locations, I have never found a situation in which the OSHA regulations did not provide a reasonable solution to problems. I know this is not a popular view.

Usually, the objections I hear to the OSHA regulations are based on ignorance. For example, I heard a commentator on a public television program complain that OSHA ridiculously requires manufacturers to provide technical data sheets on everything, including compressed air! Actually, this is true because there are four grades of compressed air, and only one is pure enough to be used for breathing apparatuses.

So far, every objection I have heard has a logical answer or a method of obtaining a variance from OSHA. In fact, my personal reservations about the regulations are that the protection they provide is too minimal, and OSHA enforcement is universally sporadic and weak.

There are, however, some very real difficulties in applying the OSHA regulations to our business. This is especially true of OSHA's training requirements.

TRAINING REQUIREMENTS

A great number of the OSHA regulations require formal training for workers, or in some cases, certificates of competence or licenses. For example, employers must provide site-specific documented training by a competent person for any worker who will wear a toxic dust mask, gloves, or protective eyewear; climb a scaffold; drive a powered lift; work on any unguarded set element that is six feet or more above the floor; be exposed to noise above certain levels; and so on.

All employers on permanent job sites are required to provide this formal training. That's how it should be. But training is never going to happen on locations and short-term jobs such as commercials. Some of these jobs only last a few days. If the OSHA training was done, those few days would be taken up with training.

Training takes time and costs money. OSHA expects the employer to pay for the training and to pay workers while they are being trained. However, some unions are doing their own training at their own expense. They are sending out workers who already have the basic information or hold the certifications. This is a great benefit to employers, who now only have to acquaint workers with conditions and equipment at the specific job site.

Graduates of university theater programs are presumed to be "trained." Unfortunately, this is rarely the case. Many theater departments are not even training their own teachers, in direct violation of the OSHA regulations. These teachers actually are unable to train their students, because they themselves do not know the regulations. Students graduating from these schools are unfamiliar with job safety regulations, their rights, and how to address the hazards they face.

But whether employers, unions, or schools do the training, this industry needs to address methods of ensuring that all workers know how to work safely and legally on the job. And currently, this isn't happening.

UNIONS

Readers also should know that, in my opinion, our business would be radically different without the unions. People today are so anxious to get into the business at almost any level that most would work for free and under any conditions. Every salary increase and benefit that is negotiated by unions sets a standard that helps both union and nonunion workers alike to make a decent living.

The unions, in my opinion, are the only check on the current trend to apply a greater proportion of the production budget to the top: the directors and the stars. This means that less money proportionally is going to the craft and technical workers and to non-star performers. I see union contract negotiations becoming more difficult, salary increases failing to keep up with living costs, and givebacks in benefits and hours. This is unconscionable, especially in a time of great prosperity.

I also see union contracts as the only reasonable alternative to producers who hire workers as independent contractors. The independent contractor status is potentially disastrous for craft and technical workers. It means these workers are unprotected by the OSHA regulations and also must provide all their own medical, pension, insurance, and other benefits. Worse, independent contractors can be held liable for accidents or legal claims related to their work.

For example, a lighting designer working for a theater as an independent contractor was recently named in a lawsuit brought by an individual claiming respiratory damage from the special effect fog chemical that he used. One of the first questions he was asked in deposition was whether or not he had liability insurance. I don't think this individual was aware until that moment that his personal assets were at risk. And he will have to pay for his own legal defense, right or wrong!

I would counsel every worker in this business to join the appropriate union and work under union contracts. And if there are things about the unions you don't like, join and change them. I know it's possible because I've seen it happen. You are your union.

USING THIS BOOK

I wish I could follow the advice of so many of my colleagues and make the material in this book simple and untechnical. This is not possible. Complex regulations, chemical products, safety equipment, ventilation, and a host of unavoidably technical subjects must be covered.

Instead, I have worked very hard to try to make these complex subjects understandable. And if you find yourself confused, I have not left you to ponder alone. You are welcome to contact me. There are two ways to do this:

1. You can reach me through a nonprofit organization I founded called Arts, Crafts and Theater Safety (ACTS). We answer an average of thirty-five inquiries per day by phone, mail, or e-mail. Answering inquiries is one of our free services. ACTS also publishes a newsletter and provides a number of technical services at below–market-value cost.
2. You can contact me through the United Scenic Artists, Local 829, International Alliance of Theatrical and Stage Employees (IATSE). I am their safety director. The United Scenic Artists have generously made it their policy to invite members of any IATSE union or, in fact, any other union in our business to contact me. All of us must work together to make workplaces safe.

Readers can contact me in either capacity at:

181 Thompson St., # 23
New York, New York 10012-2586
phone: (212) 777-0062
e-mail:ACTSNYC@cs.com
www.usa829.org
www.caseweb.com/ACTS

1 The Way It Is

Before we try to improve safety and health conditions in our business, we should spend a few minutes looking at the entrenched behavior that will need adjusting. Most readers will already be familiar with many of these attitudes and may smile in recognition of some of them.

HASTE: IT DOESN'T HAVE TO BE SAFE, IT HAS TO BE TUESDAY

Rarely, if ever, does this industry allow enough time for any phase of production, from planning to execution. Time is money, and the time spent doing things safely isn't visible from the audience like a special effect or a new backdrop is. How shortsighted it is to assume that there is no time to do a thing right, but somehow there will be time to do it over.

UNSAFE CONDITIONS: JURY-RIG AND HOPE IT HOLDS

Many venues and shops have bad ventilation systems; unsafe walking surfaces and stairs; outdated, poorly maintained, or unguarded saws and other equipment; poor work lighting; and recycled scenery.

Unsafe venues and shops should be repaired or go dark. Unsafe equipment should be repaired, replaced, or taken out of service. Productions and theater activities should be planned around a facility's limitations. For example, only shops equipped with spray booths should do spray painting; and shows should be staged with only limited amounts of scenery, props, or lighting effects if doing otherwise means using unsafe equipment.

BAD ATTITUDES

Macho: It's Alive, Well, and Living in a Theater

Performers and technicians alike commonly believe that suffering, risk-taking, and even dying for art is an appropriate price to pay for the privilege of working in the field. Actors use the stock phrase, "The show must go on." But theater arts are not the equivalent of wilderness survival expeditions; they are a part of the humanities curriculum in many universities. And in the professional theatrical and entertainment industries, no level of casualties is acceptable!

Theatrical amateurs and professionals alike are charged with enlightening and enriching audiences and themselves, not risking the life and limbs of both. When an effect or stunt is associated with a risk, the risk must be carefully assessed. Risks a professional performer takes must be minimal, calculated, and limited to a minor adverse consequence at worst. Risks to young student performers or to audiences are not acceptable at all.

Horseplay: Only for Horses and Jackasses

Breaking rehearsal tension or keeping up team spirit is not, as some assume, accomplished by permitting horseplay. Horseplay is a well-known cause of injuries and accidents. Directors, producers, supervisors, or shop stewards who permit or encourage horseplay may find themselves and/or their employers liable for any resulting accidents.

Cast Party Mentality: Damn the Strike, The Party's On

Parties must be scheduled at times that will not encourage workers to rush hazardous activities. This is particularly true of strikes at the end of runs. Strikes combine all the hazardous activities of lighting, rigging, electrical work, and construction at one time. A looming party should not tempt workers to give in to haste and the destructive urges that transform disassembly into demolition.

IMPAIRED JUDGMENT

Drugs and Alcohol: The Barrymore Mystique

You have to keep your wits about you in this business. Yet many choose to impair their wits by using drugs or alcohol recreationally. Just as these chemicals cause accidents on the highway, they are responsible for accidents in theater as well.

Chemicals in Products: The Hidden Enemy

Besides alcohol and street drugs, other chemicals can cause narcosis and impaired judgment. You may inhale these quite innocently while working with paints, costume cleaning solvents, aerosol spray products, and the like. Whether inhaled accidentally or deliberately (as in glue sniffing), no one should have these chemicals in his or her bloodstream while they work.

Medications

Some people must take medications that impair their judgment and timing, or that may interact with other chemicals they inhale while using paints and other products. People who must use prescribed medications should check with their doctors about such interactions. They also should know whether their medications will affect their ability to operate dangerous machinery, drive, or do other hazardous jobs. Teachers and supervisors should know about the physical condition and limitations of their students or workers before assigning tasks.

Lack of Sleep or Food: We'll Stop When We Drop

A common theater practice is to compensate for the lack of sufficient production time by working incredibly long hours. Overwork and lack of proper nourishment impair judgment and awareness and have caused accidents.

In school or community theater, long hours amount to a kind of endurance test given to new people to see if they "can take it" or to prove they are sufficiently "dedicated." Some teachers rationalize this hazing of students by telling them they are getting a dose of how things are in the "real world." This is usually done with a vengeance by teachers whose own aspirations have not been realized in that real world.

In any case, a world that allows people to work to a dangerous state of exhaustion should not be emulated. And the best way to correct this problem is to flood the business with students and apprentices who know how things should be done and can recognize abuse when they see it.

Psychological Stress

By nature, the theater is a psychologically difficult environment, rife with personal and artistic pressures. However, when additional pressures, such as a tyrannical director or unreasonable job insecurity, are added to this already difficult environment, stresses may be created that can hamper productions and contribute to poor judgments and accidents.

WORKING ALONE: IT'S 2 AM—DO YOU KNOW WHO IS IN YOUR VENUE?

There are pits, electrical equipment, catwalks, toxic chemicals, and a host of deadly hazards in a theater. Never allow anyone to work alone or even remain in a venue alone. Organize "buddy system" work schedules for odd hours.

UNQUALIFIED WORKERS: I'LL TRY ANYTHING ONCE

Actors are so anxious to act, and technicians are so eager to get a job, that they will often claim expertise they do not have. Producers, managers, and directors often are content to let them try. Unqualified technicians put themselves and others at risk if they use lasers, holography, pyrotechnics, fog and smoke effects, industrial plastics, paints, dyes, and other toxic materials and processes.

LAWLESSNESS: RULES AND REGS ARE FOR OTHER FOLK

Many people in our business believe that health and safety regulations and laws can be bent or broken because theater and entertainment are "special." This incorrect idea is bolstered by the fact that authorities often fail to inspect many theaters and entertainment venues and shops, so they may escape penalties for ignoring regulations. Retribution comes if an accident, fire, or incident occurs that brings a facility to authorities' attention. The liabilities and penalties resulting from ignoring the rules usually more than cancel any benefits.

2 Safety: It's the Law

Our business is not "special" when it comes to safety. The same occupational laws that apply to any business also apply to us.

OCCUPATIONAL SAFETY RULES

In the United States, the law governing the relationship between employers and employees is called the Occupational Safety and Health (OSH) Act. In Canada, the equivalent law is called the Occupational Health and Safety (OHS) Act. Other countries have similar laws. But whatever the country, the purpose of these laws is to protect workers.

For example, the OSHAct general duty clause reads in part that the "employer shall furnish . . . employment and a place of employment which are free from recognized hazards." The Canadian OHSAct requires employers and supervisors to "take every precaution reasonable in the circumstances for the protection of a worker." These brief general statements serve as the foundation for complex regulatory structures.

WHICH OSHA IS YOUR OSHA?

The first step to comply with these laws is to find out exactly which laws apply to you. For example, U.S. workers may come under either federal or state OSHA regulations. In Canada, each province and territory has its own set of rules. A copy of the applicable laws can be obtained from federal, state, or provincial departments of labor and should be kept handy for reference.

Most U.S. workers come under the federal OSHA regulations. But some states have their own federally approved state OSHA plans. The state rules usually are similar or even identical to the federal ones, but enforcement is by the state. Such states include Alaska, Arizona, California, Hawaii, Indiana, Iowa, Kentucky, Maryland, Michigan, Minnesota, Nevada, New Mexico, North

Carolina, Oregon, Puerto Rico, South Carolina, Tennessee, Utah, Vermont, Virginia, Virgin Islands, Washington, and Wyoming.

The federally approved state OSHA rules must change whenever new federal standards are adopted. States must adopt comparable standards within six months of the publication date of new federal standards. The protection afforded workers under state rules must be equal to, or better than, the federal standard.

Some states exempt their own state, county, and/or municipal workers from the federal laws. Some of these workers, such as those in New Jersey and Massachusetts, are protected by very strong public employee laws. Other states, which will remain nameless, have poor state standards and/or extremely weak enforcement. Some of these poorly protected employees work in theater and film departments at state universities and colleges, or in state, county, and civic theaters.

This book obviously cannot cover all the different state and public employee protection laws and their problems. *So in all cases, the U.S. federal rules will be cited in this text.*

TWO SETS OF LAWS

Whether the applicable regulations are state or federal, the laws are further divided into two sets of regulations: rules for general industry and rules for construction work.

Workers in a permanent venue, such as a theater or an established scene shop, are protected under the General Industry Standards (29 CFR 1900-1910). Work on location, where the venue is temporary, is usually regulated under the Construction Industry Standard (29 CFR 1926). However, OSHA's definition of construction work is broad and includes any "alterations or repair, including painting and decorating." This means that any large set construction in shops, on stage, or on location is regulated under the construction standard. Both standards will be mentioned in this text when appropriate.

APPLICABLE OSHA REGULATIONS

The OSHA regulations apply to every inch of a workplace. While all the rules should be followed, the ones listed below are particularly relevant to our business.

Hazard Communication (1926.59, 1910.1200)

This law provides protection for any worker who may be exposed to potentially harmful substances on the job, such as paints, solvents, dyes, wood dust, and welding fumes and other products associated with scene, prop, and costume shops.

Other examples of covered chemical products might include: solder and touch-up paints in lighting shops; spray products and solvents used in wig and makeup departments; diazo copiers, photocopy machines, and solvent-based markers in design studios and offices; and special effects emissions, such as those from theatrical fog, haze, smoke, or pyrotechnics.

The Hazard Communication Standard is the cornerstone of a facility's health and safety programs. It requires employers to develop a program to inform and train all full- and part-time employees who may be exposed to toxic substances. (Sections of this book can be used as training materials for hazard communication.) The law requires:

- *A written hazard communication program.* Employers must write a plan describing precisely how they will meet each provision of this law. In 1999, OSHA cited employers more often for failure to have a written program than for any other violation. OSHA inspectors are likely to review this program first, because without it, other provisions of this law cannot be properly instituted. (Sources for prototype programs are in the Bibliography.)
- *An inventory.* All potentially hazardous products must be listed and their location in the facility indicated.
- *Material Safety Data Sheets (MSDSs).* The MSDSs for all potentially hazardous materials must be filed in a manner that can be easily retrieved (see chapter 6).
- *Labels.* All containers of chemicals must be labeled with the name of the substance, any required hazard warnings, and the name and address of the manufacturer. This also applies to products that are bought in large containers and transferred to smaller ones.
- *Training records.* There must be documentation that training by a qualified person has been given to all employees who may be exposed to toxic chemicals and that the employees comprehended the training.
- *Provision for ready access.* The MSDSs and all written elements of the program must be made available to workers during all working hours.

Respiratory Protection (1926.103, 1910.134)

Employers whose workers use respirators of any type from half-face masks to supplied air-breathing apparatuses must establish a respirator program. This rule requires a written program, a check on the employees' medical status to ensure that they are physically able to tolerate the breathing stress induced by masks and respirators, professional fit testing of workers, procedures for equipment maintenance, and training of workers. (See chapter 8 for details.)

Personal Protective Equipment (1926.28,1910.132-133)

OSHA now requires formal, documented training in the proper use of gloves, safety glasses, aprons, and other protective equipment to prevent untrained workers from using the wrong equipment. For example, I have often seen scenic artists wearing chemical splash goggles when using a grind wheel. Splash goggles will not protect them from the impact created by a flying chip.

Ladders (1910.25-26, 1926.1053)

Both wooden and metal ladders of various types must be used for specific purposes and tasks at specific heights. All theater workers must be familiar with these rules.

Fall Protection (1926.500-503, 1910.23)

If people have to work on a set from which they could fall more than six feet, or if they work on a scaffold from which they could fall ten feet, the elevation must be standard-railed or the worker must be tied off and harnessed. For shops that are in permanent locations, the height at which guarding is required is four feet. OSHA requires documented training for all workers who use fall protection equipment.

Scaffold Regulations (1910.28 and 1926.451)

These rules have recently changed. There now must be a "competent person" who directs erection and use of scaffolds. Training in the proper use and rules is required for workers who work on scaffolds and for the competent person.

Powered Industrial Lifts (1926.178)

This rule requires operators of forklifts, platform lift trucks, and other powered lifts, including Genie truck lifts, to be trained before they operate them. Training must consist of both classroom and practical training in proper vehicle operation, the associated hazards, and requirements of the OSHA standard for powered industrial trucks.

Emergency Plans and Fire Prevention (1910.35-38, 1926.150, 155-166)

Employers must have written emergency plans, and all workers must be trained at least annually to know what the alarms or other warnings mean, where to exit, where to meet for a head count, and so on. If workers are expected to use a fire extinguisher, they must be trained about the equipment, when to fight and when to flee, etc.

Medical Services and First Aid (1910.151, 1926.50)

Programs for dealing with medical emergencies must be in place. Workers who are expected to provide CPR, first aid, or other services must be trained and certified. First aid kits should be readily available. Eyewash stations and emergency showers must be within a short distance (as defined by the standard) of any use of irritant or corrosive chemicals, such as oil-based paints, plastic resin systems, or solvents.

Bloodborne Pathogens Standard (1910.1030)

This standard protects workers from exposure to blood and body fluids. Its requirements include formal training, protective equipment, and special containers for disposal for contaminated items, such as gloves, bandages, or sharp items that have punctured the skin. A court case established that the standard is applicable to workers in the textile industry who use tagging guns and needles. Clearly, this law covers costume shop workers, carpenters, and any other workers who routinely suffer small cuts and accidents.

Occupational Noise Exposure (1910.95 or 1926.52)

This law applies to all loud sound, from machinery to music. The sound levels must be measured and records kept of the decibels to which workers are exposed. If the noise levels are at or near the permissible noise limit, hearing protection is required. The employer must provide workers with audiograms to test their hearing acuity, and they must train workers about hearing loss and the use of protective devices.

Lead in Construction (1926.62)

Before any old paint (applied prior to 1980) can be removed or even disturbed, it must be analyzed for its lead content. If it is lead-based, the employer must use trained lead-abatement workers. All workers must be informed about the presence of lead paint or other sources of lead in the work environment. (See also Location Hazards, chapter 19.)

Lead in General Industry (1910.1025)

Once it is known that there is lead in the environment or that lead is being used (e.g., soldering or using lead pigments), the employer must do personal monitoring of those employees who may be exposed during work. If these tests show airborne lead above the standards action level, then many other expensive provisions take effect, including blood-lead tests, showers, and changing rooms. If the tests are below the action level, there still must be complete record keeping and retesting every time conditions, jobs, or personnel change.

Asbestos in Construction (1926.1011)

The asbestos-in-buildings rule requires employers in old buildings to have an asbestos management plan to which workers have access on request. This plan should tell workers exactly which insulation, ceiling tiles, vinyl floor tiles, etc. are asbestos-containing. All insulated asbestos pipe must be clearly labeled. (See chapter 20, Asbestos.)

Electrical Safety (1926.401-405, 1910.301-333)

The federal OSHA standard references the National Electric Code (NEC). This means that any wiring or appliance that does not meet the NEC code is also an OSHA violation. Common violations seen in shops and locations include non-compliant panels and cabinets, panels without clear access in front of them (1910.302(g)(1)(i)), plugs that have been altered by clipping off the ground, equipment that is not either ground faulted or double insulated, and outlets and extension cords that are not ground fault circuit interrupted (1926.404(b)(1)).

Hazardous Waste Operations/Emergency Response (1910.120)

These rules also require training for any person expected to deal with disposal of paints, solvents, or other hazardous waste. (There are also Environmental Protection Agency [EPA] rules that apply to storage, handling, and shipping of waste.)

Flammable and Combustible Liquids (1910.106 or 1926.152)

Solvent-containing paints, spray cans, and other flammable products must be properly stored (e.g., in nonflammable storage cabinets) and dispensed.

Housekeeping (1910.22, 1926.25)

All places of employment, passageways, storerooms, and service rooms shall be kept clean and orderly and in a sanitary condition. The floor of every workroom shall be maintained in a clean and, so far as possible, dry condition (e.g., drainage of wet areas by false floors, use of platforms, mats, or other dry standing places).

To facilitate cleaning, every floor, working place, and passageway must be kept free from protruding nails, splinters, holes, loose boards, or any materials over which people may trip. Aisles and passageways must be maintained and kept clear, even during set construction, loading in, or installation.

Sanitation (1910.141)

This rule applies to bathrooms and general sanitation. Two particular provisions in (g)(2) and (g)(4) prohibit employees from eating or storing food in areas where toxic substances are used or stored. Eating lunch or drinking beverages must be done in a sanitary room, complete with walls, doors, and a separate ventilation system to isolate it from work areas.

Machinery Rules

There are a number of standards that apply to specific types of machinery. Included are rules that require guards at the point of operation and other very precise precautions for specific types of woodworking machinery (1910.213), abrasive grind wheels (1910.215), and drive belts and flywheels (1910.219). Similar safety and guarding rules apply to handheld and portable power tools (1910.242).

Welding, Cutting, and Brazing (1910.251-255, 1926.350-351)

Welding done in theaters and shops comes under many regulations—not only the standards above, but rules about compressed gas cylinders, electrical hazards (arc), etc. A common violation seen in theaters and shops is welding occurring less than thirty-five feet from combustible materials, such as wood dust or styrofoam, and on plywood floors with cracks between the sections (1910.252(a)).

Confined Space (1910.146)

This complex rule applies to any container, room, or space large enough to hold a person, but which has only one entrance/exit. Air contaminants can build up in these places quickly and become life-threatening. These spaces are divided into two types: permitted and nonpermitted. I have seen storage areas and crawl spaces in old theaters and stadia that qualify for this rule. Such spaces require worker training and notification.

Biological Hazards

Theaters and film locations are often old buildings that have sanitation problems. Animal droppings or nests from pigeons and rodents, water-damaged or moldy materials, and other disease-carrying substances are "recognized hazards" and, as such, can be cited under the OSHA General Duty Clause (Section 5(a)(1)). Cleanup of these materials may require professional abatement or training of regular workers in use of protective equipment and disinfectants.

COMPLIANCE

Ordinarily, a discussion of compliance with OSHA regulations could consist of two words: "do it!" However, it is not that simple in our business.

While many employers do a good job of complying with OSHA regulations, it has been my experience that the majority do not. I personally know of a major TV network studio that has been obligated by law since 1985 to provide hazard communication training for its scenic artists and has refused to do it, even when it was offered by the union for free. And at five film locations, I have provided the documentation needed to stop work because the employers were exposing workers to lead, asbestos, and other serious hazards.

One of the reasons for general noncompliance is that production budgets do not allocate the money for safety. This is not acceptable, since the preamble of every OSHA rule includes an estimated cost per employee for both initial compliance and program maintenance. Initial compliance, during which the program is set up, is always most costly. Program maintenance is much cheaper.

As each new rule has been published, OSHA has expected employers to spend this money. Yet many theater, film, and university theater program budgets have not reflected these costs for decades. Now, employers are often stunned when they look at the total cost of the number of regulations, such as those listed above, with which they must comply. This cost would have been very manageable had they complied with each regulation when it was new.

Catching up with the cost of compliance is going to be difficult for many employers. And I say this with apologies to the smaller number of wonderful employers that have professional safety people on their staffs and their programs in place. They are the models that prove that following the rules is possible.

3 Our Materials: How They Affect Us

Paints, plastics, wood dust, hair spray, stage fog, and most of the products we use have hazardous ingredients. Information about these chemical ingredients can be found in occupational health literature and on data sheets provided by manufacturers. But in order for us to understand the information in the literature, we must go over the basic terminology.

BASIC CONCEPTS

Dose

Chemical toxicity is dependent on the dose—that is, the amount of the chemical that enters the body. Each chemical produces harm at a different dose. Highly toxic chemicals cause serious damage in very small amounts. Moderately and slightly toxic substances are toxic at relatively higher doses. Even substances considered nontoxic can be harmful if the exposure is great enough.

Time

Chemical toxicity is also dependent on the length of time over which exposures occur. The effects of short and long periods of exposure differ.

Short-Term or Acute Effects

Acute illnesses are caused by large doses of toxic substances delivered in a short period of time. The symptoms usually occur during or shortly after exposure and last a short time. Depending on the dose, the outcome can vary from complete recovery, through recovery with some level of disability, to—at worst—death. Acute illnesses are the easiest to diagnose, because their cause and effect are easily linked. For example, exposure to solvents in oil-based paints can cause effects from lightheadedness to more severe effects, such as headache, nausea, and loss of coordination. At even higher doses, unconsciousness and death could result.

Long-Term or Chronic Effects

These effects are caused by repeated, low-dose exposures over many months or years. They are the most difficult to diagnose. Usually, the symptoms are hardly noticeable until severe permanent damage has occurred. Symptoms appear very slowly, may vary from person to person, and may mimic other illnesses. For instance, chronic exposure to oil-based paints during a lifetime of painting may produce dermatitis in some individuals, chronic liver or kidney effects in others, and nervous system damage in still others.

Cumulative/Noncumulative Toxins

Every chemical is eliminated from the body at a different rate. Cumulative toxins, such as lead, are substances that are eliminated slowly. Repeated exposure will cause them to accumulate in the body.

Noncumulative toxins, like alcohol and other solvents, leave the body very quickly. Medical tests can detect their presence only for a short time after exposure. Although they leave the body, the damage they cause may be permanent and accumulate over time.

The Total Body Burden

This is the total amount of a chemical present in the body from all sources. For example, we all have body burdens of lead from air, water, and food contamination. Working with lead-containing art materials can add to this body burden.

Multiple Exposures

We are carrying body burdens of many chemicals and are often exposed to more than one chemical at a time. Sometimes, these chemicals interact in the body.

Additive effects occur when one chemical contributes to or adds to the toxic effects of the other. This can occur when both chemicals affect the body in similar ways. Working with paint thinner and drinking alcohol is an example.

Synergistic effects occur when two chemicals produce an effect greater than the total effects of each alone. Alcohol and barbiturates or smoking and asbestos are common examples. In fact, smokers are at far greater risk in general of developing cancer and other diseases of the lungs.

Mutations

These can be caused by chemicals that alter the genetic blueprint (DNA) of cells. Once altered, such cells usually die. Those few that survive will replicate themselves in a new form. Any body cell (muscle, skin, etc.) can mutate.

Cancer. When the mutated cells are capable of reproducing rapidly enough to become invasive, they are cancer cells. Chemicals that cause cancer are called carcinogens. Examples include asbestos and benzidine dyes and pigments. Unlike ordinary toxic substances, the effects of carcinogens are not strictly dependent on dose. No level of exposure is considered safe. However, the lower the dose, the lower the risk of developing cancer. For this reason, exposure to carcinogens should be avoided altogether or kept as low as possible.

Occupational cancers typically occur ten to forty years after exposure. This period of time, during which there are no symptoms, is referred to as a latency period. Latency usually makes diagnosis of occupational cancers very difficult.

Genetic Damage. When the cell that mutates is a human egg or sperm cell, mutagenicity can affect future generations. Most pregnancies resulting from mutated sperm and eggs will result in spontaneous termination of the pregnancy (reabsorption, miscarriage, etc.). In other cases, inherited abnormalities may result in the offspring.

Birth Defects. Chemicals that affect fetal organ development cause birth defects. These chemicals are called teratogens. They are hazardous primarily during the first trimester. Two proven human teratogens include the drug thalidomide and grain alcohol. Chemicals that are known to cause birth defects in animals are considered "suspect teratogens." Among these are many solvents, lead, and other metals.

Fetal Toxicity

Toxic chemicals can affect the growth and development of the fetus at any stage of development.

Allergies

Allergies are adverse reactions of the body's immune system. Common symptoms may include dermatitis, hay fever, and asthma. Although a particular person can be allergic to almost anything, certain chemicals produce allergic responses in large numbers of people. Such chemicals are called "sensitizers." Well-known sensitizers include epoxy adhesives, some dyes, and wood dust.

The longer you work with a sensitizing chemical, the greater the probability you will become allergic to it. Once developed, allergies tend to last a lifetime, and symptoms may increase in severity with continued exposure. A few people even become highly sensitized—that is, develop life-threatening reactions to exceedingly small doses. This effect is caused by bee venom in some people, but chemicals such as rubber latex and urethane foaming products have produced similar effects, including death.

HOW CHEMICALS ENTER THE BODY (ROUTES OF ENTRY)

In order to cause damage, toxic materials must enter your body. Entry can occur in the following ways:

Skin Contact

The skin's barrier of waxes, oils, and dead cells can be destroyed by chemicals such as acids, caustics, solvents, and the like. Once the skin's defenses are breached, some of these chemicals can damage the skin itself, the tissues beneath the skin, or even enter the blood, where they can be transported throughout the body, causing damage to other organs.

Cuts, abrasions, burns, rashes, and other violations of the skin's barrier can allow chemicals to penetrate into the blood and be transported throughout the body. There are also many chemicals that can—without your knowing it—enter the blood through undamaged skin. Among these are wood alcohol and benzene.

Inhalation

Inhaled substances are capable of damaging the respiratory system acutely or chronically at any location, from the nose and sinuses to the lungs. Examples of substances that can cause chronic and acute respiratory damage include acid vapors and fumes from heating or burning plastics.

Some toxic substances are absorbed by the lungs and are transported via the blood to other organs. For example, lead from soldering may be carried via the blood to damage the brain and kidneys.

Ingestion

You can accidentally ingest toxic materials by eating, smoking, or drinking while working, pointing brushes with your lips, touching soiled hands to your mouth, biting your nails, and similar habits. The lung's mucus also traps dusts and removes them by transporting them to your esophagus, where they are swallowed.

Accidental ingestion can occur when people pour chemicals into paper cups or glasses and later mistake them for beverages. Some of these accidents have even killed children.

ARE YOU AT RISK?

Everybody is susceptible to occupational exposures. How they are affected will depend on the nature of the chemical, its route of entry, the degree of exposure, and, in some cases, the susceptibility of the exposed person. For example, most people's lungs will be affected equally by exposure to similar concentrations of strong acid vapors. However, someone who already has bronchitis or emphysema may be even more seriously harmed.

How Much Is Too Much?

Clearly, most of us work with some toxic materials. But how serious is this exposure? At what point should you become concerned? To help you (or your doctor) find out if your exposure is significant, you should be able to answer the following questions about each of the materials you use:

- *How much do you use?* Keep track of the amounts of materials you use. Obviously, a gallon of a substance is potentially more hazardous than a pint.
- *Under what conditions are you exposed?* Does your workplace have good ventilation and protective equipment, or are you working in an unvented space, breathing the vapors, getting it on your hands, eating from a table set up in the workplace, etc.?
- *How often are you exposed?* Do you use the material every day, twice a week, or once a month? Do you use it two hours a day, eight hours a day, or longer? Since we often have varied work schedules, you may need to keep a work diary in order to answer this question.
- *How toxic are the materials you use?* Some materials are much more toxic than others. Learn which materials you use are the most hazardous. (See chapters 5, Labels, and 6, Material Safety Data Sheets.)
- *What are your "total body burdens" of toxic substances?* Your total body burden of a substance is the total amount of that substance in your body from all possible sources. For instance, if you work with paints all day and also paint for a hobby, your total body burden comprises your exposure both at work and at home. For another example, alcoholic beverage consumption adds to the effects of solvent exposure on the job.
- *Are you a member of a high-risk group?* Some people are physiologically much more susceptible than most to the harmful effects of chemicals. Among these are smokers, drinkers, children, the elderly, people with chronic diseases or allergies, and people taking certain kinds of medications. At especially high risk is the pregnant woman and her baby. All of these groups should be especially vigilant in avoiding workplace chemical exposures.
- *Are you developing physical symptoms that are not easily explained?* It is time for you to evaluate your materials to see if they are known to cause these symptoms. Some of the illnesses and symptoms caused by materials we use are listed below. People taking certain medications are sometimes at higher risk. For example, medications (and recreational drugs) that are narcotic may potentiate the effects of solvents and metals that also are narcotic (i.e., affect the brain).

OCCUPATIONAL ILLNESSES

Any organ in the body can be affected by an occupational illness, including the following:

- *The Skin.* Many chemicals can affect the skin by causing irritation, damage to the outer layers, or by causing allergic skin reactions called allergic dermatitis. Other occupational skin diseases include infections and skin cancer. Working with cuts, abrasions, or other kinds of skin damage may lead to infections. Lampblack pigments and ultraviolet light are associated with skin cancer.
- *The Eye.* Chemicals that are irritating or corrosive (e.g., acids) can damage the eye severely. Ultraviolet light (e.g., from welding) can cause eye damage. A few chemicals, such as methanol (in some shellacs) and hexane (in some rubber cements) can also damage eyesight when inhaled or ingested.
- *The Respiratory System.* People were designed to breathe air. Almost any inhaled substance can cause respiratory problems. Symptoms can range from minor irritation to life-threatening types of pneumonia or anaphylactic shock. Exposures to chemicals over years can cause chronic respiratory damage, such as chronic bronchitis, lung scarring (e.g., from asbestos or silica), and cancer. Often, the first symptom of respiratory problems is an increased susceptibility to colds and respiratory infections.
- *The Heart and Blood.* Many solvents at high doses can alter the heart's rhythm (arrhythmia) and even cause a heart attack. Benzene, still found as a contaminant in some solvents and gasoline, can cause aplastic anemia (decreased bone marrow production of all blood cells) and leukemia.
- *The Nervous System.* Metals like lead and mercury, and almost all solvents, can affect the nervous system. Symptoms can vary from mild narcosis (lightheadedness, headache, dizziness) to coma and death at high doses. Chronic exposure can cause short-term memory loss, mental confusion, sleep disturbances, hand-eye coordination difficulties, and depression. Some chemicals such as n-hexane (found in rubber cements, aerosol sprays, etc.) can damage the nervous system in ways that can result in a disease similar to multiple sclerosis.
- *The Liver.* Hepatitis can be caused by chemicals as well as by disease organisms. Some toxic metals and nearly all solvents, including grain alcohol, can damage the liver if the dose is high enough. Liver cancer is caused by chemicals such as carbon tetrachloride.

- *The Kidney.* Kidney damage is also caused by many metals and solvents. Lead and chlorinated hydrocarbon solvents such as trichloroethylene are particularly damaging. Heat stress and accidents (damaged blood and muscle cell debris can block kidney tubules) are also causes of kidney damage.
- *The Bladder.* Benzidine-derived pigments and dyes are documented causes of bladder cancer (see chapter 12).
- *The Reproductive Effects.* Chemicals can affect any stage of reproduction: sexual performance, the menstrual cycle, sperm generation, all stages in organ formation and fetal growth, the health of the woman during pregnancy, and the newborn infant through chemicals secreted in breast milk. These complex effects are covered in more detail in the next chapter.

4 Reproductive Hazards

The most "creative" job anyone, male or female, can do is to produce a healthy, happy member of the next generation. And no undertaking creates more emotional and psychological changes in both men and women. People who have never given a thought to being overworked and overexposed to chemicals on the job suddenly realize that a tiny, fragile life is depending on the health of their bodies. Many of the positive changes in the workplace have been stimulated by prospective parents who began to look at our workplaces from this new perspective.

WHAT DO WE KNOW?

The National Institute of Environmental Health Science's research indicates that 5 to 10 percent of couples who want to have children are infertile; about half of all pregnancies are not successfully completed; 3 to 5 percent of newborns have major birth defects; and sperm counts may have declined in recent decades. Studies are underway to try to evaluate how many of these effects are caused by chemicals, such as those we use on the job.

A number of the chemicals used in the scenic arts, costume making, carpentry, welding, and prop making are known or suspected of being able to affect human reproduction. Most of us try to avoid chemical exposures when we start planning a family, but just how hazardous our chemical products are to pregnant theater artists is unknown.

A search of the literature produced no studies of theater, film, and TV workers' reproductive problems. There are, however, studies of industrial workers who are exposed to the same chemicals that we use. For example, industrial painters, sign makers, and printers are all exposed to pigments and solvents. Workers in the textile industry are exposed to dyes and fabric dusts. Construction workers inhale wood dust, plaster, and other dusts from building materials. Welders inhale metal fumes.

We can expect, then, that scenic artists and industrial workers using the same materials will experience the same reproductive health problems. In fact, theater artists may be at greater risk than industrial workers if they toil longer than eight hours a day, work without proper ventilation and protective equipment, or, worst of all, also work at home, where intimate and prolonged exposure can occur.

PREGNANCY

During pregnancy, the fetus can sustain two types of damage from exposure to toxic substances: birth defects and toxic effects.

Birth Defects

Birth defects can be caused by chemicals or drugs that alter the development of organs. This means that birth defects only occur in the first three months of pregnancy, as the organs are forming. For example, the drug thalidomide affects the limbs only during formation. Once they are fully formed, the drug has no effect.

Toxic Effects

These are a kind of "poisoning" that can occur at any stage of pregnancy and even after birth. For example, lead can damage brain function at any time from early in conception even through adulthood. But it is the fetus that is most susceptible.

CHEMICALS IN OUR MATERIALS

A vast array of chemicals are used in scene, costume, prop, electrical, carpentry, and other shops. There are over two thousand dyes and about three hundred pigments commercially available. Over forty-five different metals are found in pigments, bronzing powders, welding, brazing, and soldering alloys, and similar craft materials. Hundreds of chemical solvents can be used to thin hundreds of different natural and synthetic resins, oils, and waxes in various paints, inks, varnishes, glues, adhesives, and fixatives. If we count the chemical additives in use to modify paints, plastic resin products, and these other products, then the numbers of chemicals scenic artists use number tens of thousands.

Only a tiny fraction of these substances have been studied for reproductive effects. Our failure to require chemicals to be tested for long-term hazards essentially makes humans an experimental species. For example, a study showed that one third of pregnant IBM computer chip workers in a class called "glycol ethers" miscarried after exposure to very small amounts of chemicals. These were chemicals in common use at the time. They have been used since the 1970s in many spray cleaners, latex paints, water-based printmaking inks, liquid dyes, felt-tip pens, and spray paints.

Today, those same products may still contain either the same glycol ethers or ones that are closely related to them. Until more is known, it is wise to avoid unnecessary exposure to these and many other materials. Two types of chemicals in particular should be avoided: solvents and metals.

Solvents

The term "solvent" is applied to organic chemical liquids that dry faster than water and are used in products such as paints, inks, cleaners, paint strippers, aerosol sprays, marking pens, and the like. The solvent for which we have the most data is ethyl alcohol, since we also drink it in alcoholic beverages. Alcohol consumption causes almost all types of adverse reproductive effects including:

- Reduced male sex drive and performance
- Reduced male and female fertility
- Spontaneous abortion
- Birth defects
- Growth retardation and functional deficits in the fetus
- Breast milk contamination

It doesn't matter whether alcohol or other solvents are absorbed into the body by drinking them or inhaling them. While it may take a number of drinks of alcohol to cause effects, it takes much less inhalation of other, more toxic solvents to cause harm. All solvents, including alcohol, have one common characteristic: They are narcotics. "Glue sniffers" have proven that they can get high—and even die—from inhaling vapors from any solvent-containing product: glue, gasoline, or spray paints.

When the pregnant woman is exposed, the narcotic effect may damage the fetus's nervous system. Damage from drinking alcohol during pregnancy ranges from minor learning difficulties to severe facial deformities and mental retardation, depending on the amount to which the fetus is exposed. Now, it seems likely that many solvents are capable of causing similar effects.

Currently, the first study showing a connection between solvents and birth defects in humans has been published. The researchers studied Canadian women exposed to organic solvents who were employed as factory workers, laboratory technicians, artists or graphic designers, printing industry workers, chemists, painters, office workers, car cleaners, veterinary technicians, funeral home employees, carpenters, and social workers. The authors claim the study indicates that ". . . women exposed occupationally to organic solvents had a thirteen-fold risk of major malformations as well as increased risk for miscarriages in previous pregnancies" (*Journal of the American Medical Association,* March 1999).

Metals

Metals and their compounds abound in scenic materials—in pigments (cadmium red, chrome yellow, manganese blue, etc.), enamels, bronzing powders, solders, and more. Metals can be considered either as "minerals" needed for health (e.g., zinc, calcium, or iron) or as "heavy" or "toxic" metals that should be avoided (e.g., lead, cadmium, or arsenic). Actually, the toxicity of many metals lies in between these two extremes. For example, chrome and cobalt are needed by the body in tiny amounts, but they are toxic and possibly even cancer-causing in larger amounts.

TABLE 4.1 FOOD AND DRUG ADMINISTRATION (FDA)

REFERENCE DAILY INTAKES (RDIs)

MINERAL	RDIs in milligrams
calcium	1000
phosphorus	1000
magnesium	400
zinc	15
manganese	2
copper	2
chromium*	0.12
molybdenum*	0.075
selenium*	0.07

*Product labels list these in micrograms

Good reproductive health requires that we ingest a proper balance of only those metals needed by the body. This is impossible if we are supplementing our mineral intake with dusts and fumes from our studios and shops.

Lead: A Special Metal. Lead is especially hazardous. It can interfere with almost every phase of male and female reproduction. Adverse effects from lead can occur even after the child is born. Children whose blood lead content is at levels previously thought safe (10 micrograms per deciliter [μg/dL]) suffer a measurable loss of mental acuity. The younger the child, the more destructive lead is to mental function. And most vulnerable of all is the fetus.

Blood lead levels of 10 µg/dL are considered safe in adults. But if this adult is pregnant, the fetus's blood lead level is near that of the mother. This does not mean that the child born to such a woman will be retarded. Instead, it means that experts suspect that children born to these women may be less intelligent than they would have been if they had not been affected by lead.

Women who have a low blood lead count should still consult their doctors if they have had heavy lead exposure in the past. Lead stored in bones and tissues from previous exposures as far back as childhood reenters the bloodstream during pregnancy. Some physicians increase calcium supplements in such patients to reduce the amount of lead taken up by the fetus.

Mercury in Herbal Products. To protect consumers from exposure to mercury, the U.S. Food and Drug Administration (FDA) has banned mercurial drugs and antiseptics such as Mercurochrome. They also have banned mercurial soaps, makeup, and skin bleaching preparations. (The FDA does allow a tiny amount of mercury [0.0065 percent] in eye makeup as a preservative.)

However, the FDA has no jurisdiction over "dietary ingredients," as defined in the new Food and Drug Administration Modernization Act. The FDA is concerned about toxic substances, especially mercury, in herbal and natural dietary products. The FDA notes that:

> . . . mercury-containing compounds are used in traditional Chinese medicines. The Chinese *Herbal Materia Medica* reports that cinnabar (mercuric sulfide; cinnabaris, or zhu sha in Mandarin Chinese) and calomel (mercurous chloride; calomelas or qing fen in Mandarin Chinese) have been widely used as a sedative and detoxicant and to treat constipation and edema.

The extent to which mercury is actually used in these products was studied by the California Department of Health Services in 1998. The FDA reports that the study found:

> . . . that 5 of 260 traditional Chinese medicines available in the retail marketplace, which they examined, listed cinnabar as an ingredient on the label. In this study, 35 of 251 products that were screened for mercury content were found to contain significant quantities of mercury. . . . Most of the products that contained significant quantities of mercury did not list mercury sources on the label (63 FR 68775-7, December 14, 1998).

Clearly, these products pose a serious risk to consumers. Here, we have only looked at mercury, but lead and other toxic metals have been found in some of these unregulated products. In addition, some of the pure herbs themselves are hazardous to the fetus and to your health. People planning pregnancies should be very careful about the types of medications and herbal products they use.

Estrogenic Chemicals

Some chemicals mimic the effects of the female hormone estrogen. They have been shown to cause adverse male reproductive effects and birth defects in birds, fish, and mammals. Now some experts think humans are showing similar effects including:

- Worldwide reduced sperm count
- A three- to fourfold increase in cancer of the testicles
- Increases in male reproductive organ birth defects
- Decline in the ratio of male to female births in the United States, Canada, and two other countries

According to some experts, male children are now conceived and born into a virtual "sea of estrogens." The chemicals also may be affecting women by increasing rates of breast and other cancers. Substances artists may encounter that either mimic estrogen or alter hormone function include:

- Bisphenol A in some epoxy resins, other plastics, and in flame retardants
- Dioxins and PCBs, which may contaminate some dyes and pigments such as phthalo blues and greens, diarylide or benzidine yellows and oranges, and dioxazine violet
- Nonyl phenol, octyl phenol, and their derivatives, found in epoxy resins, in some latex paints, and certain detergents
- Tung oil, found in many varnishes, coatings, and inks

All of these sources can be easily avoided by wearing gloves and providing ventilation.

Other Chemicals

There are many other chemicals used in scenic work that may have reproductive effects, such as carbon monoxide, which is emitted by vehicles and forklifts in the shops. Intimate contact with organic chemical pigments and dyes should also be avoided. Many are chemically related to known carcinogens, and a few are contaminated with highly toxic chemicals such as dioxins and PCBs. These contaminants are not only toxic, they also may affect reproduction by mimicking estrogen.

COMMON SENSE PRECAUTIONS

Scenic artists can reduce their reproductive risk from toxic art materials by following some simple rules:

- *Know your materials.* Read labels and obtain material safety data sheets from the manufacturer to identify the hazardous ingredients in all your products.
- *Get advice.* Your personal physician may be a good source of information about chemicals, but the doctors who are most qualified to provide this information are board certified in occupational medicine or toxicology. Other good sources are the Local 829 United Scenic Artist's Safety Officer, (212) 777-0062, and the Pregnancy Environmental Hotline, (617) 466-8474.
- *Avoid exposure.* If there is no FDA "reference daily intake" (RDI) for a chemical with which you work, don't let it get into your body. Only take those dietary supplements approved by your doctor.
- *Listen to your body and your mind.* If a chemical makes you feel sick or "woozy" (e.g., a narcotic solvent), assume it is not good for your baby. But always keep in mind: Chemicals in amounts whose effects are not detectable in your body can, nevertheless, damage the fetus.
- *Protect yourself.* Wear gloves or use methods of working that keep products off the skin. Never eat, smoke, drink, apply cosmetics, or do any hygiene tasks in the shop. Never spray, airbrush, sand, work with dry powders, or use any material in a form that can be inhaled unless you have proper ventilation or protective equipment.
- *Ask your doctor about respiratory protection.* Occupational physicians often do not recommend respirators for people with certain health problems (e.g., heart and lung deficiencies) or for pregnant women, due to the increased breathing stress they cause.
- *Avoid lead in any form.* Never remove lead paint to prepare a wall. It is against the law anyway. Avoid lead pigments and lead-pigmented artist's paints. If you must use lead in some form, have regular blood lead tests. If you have had heavy lead exposure in years past, tell your doctor.

- *The FDA suggests that pregnant women avoid regular use of ceramicware or lead crystal glasses and mugs.* Lead is leached from many types of dinnerware—even some still sold today. Metals other than lead can leach from ceramicware, especially handcrafted ware. Use plastic or nonleaded glassware.
- *Keep children out of studios, shops, and away from toxic materials.* Exposures to some toxic substances in childhood can affect the next generation.
- *Don't dwell on past chemical exposures or exposures in the present that are too small to be significant.* What's done is done. Stress is not good for you or the baby either. If you feel worried and guilty: Welcome to parenthood!

5 Labels

To identify hazardous materials in your inventory, begin by reading product labels. While label information alone should never be considered adequate, it is your first line of defense.

LABELS AND THE LAW

In the workplace, the OSHA hazard communication standard dictates the contents of product labels. Product labels require:

- The identification of any hazardous ingredients
- Appropriate hazard warnings
- The name and address of the chemical manufacturer, importer, or other responsible party

Most products already come to the workplace properly labeled. It is when we transfer materials to other containers that we get into trouble with this law.

Transferring a material such as a paint or solvent from a large container to a bucket for individual use is not a violation of the law as long as the person making the transfer uses up the material during the course of that particular shift. But if the transferred material remains in an unlabeled container longer than this, the law has been broken. In this case, the container must be relabeled with all the required information.

The purpose of this law is very simple: There must be no "mystery" containers in the workplace. No worker should use a material unless all the required information is available on the label. And if there is an accident, illness, or fire, the emergency responders should know immediately what has spilled, caused an illness, or caught fire. For example, a container labeled only "blue paint" is not properly labeled. And containers whose contents have dripped over the label, obscuring the wording, are also in violation of the rule.

TYPES OF LABELS

There are three types of labeling that craftspeople and designers will see. These are the general consumer label, the art material label, and the industrial/professional product label. Different laws govern each type. For example, consumer labels require information on how to use the product safely.

TERMINOLOGY: READING BETWEEN THE LIES

Some of the terminology used in the three types of labels vary in their meanings and make interpreting them very confusing to the average person. In fact, many label terms are routinely misunderstood by consumers and workers. Some of these include words like: "water-based," "nontoxic," "biodegradable," and "natural." The following redefines these terms more realistically.

Biodegradable

Disposal of products bearing the "biodegradable" label is easy. Liquids may go down the drain. Solids may go out with the trash. However, do not assume that what is safer for the environment is necessarily safer for you. For example, the new refrigerants and spray-can propellants are safer for the ozone layer, but most are more toxic to people.

And remember the old phosphate detergents? They were not very toxic, and they were readily taken up by plants as fertilizer. It was because we poured *too much* detergent "fertilizer" into lakes and streams that they were banned.

We replaced the phosphates with the new biodegradable detergents and enzyme cleaners. Some of the enzymes cause serious allergies and health effects in some people. And while the detergents do not break down into simple substances that are used as fertilizer by plants, they do break down into something! Currently, scientists are investigating the possibility that degradation products from certain detergents are causing deformity in the sex organs of fish and aquatic life in wetlands.

If this were not enough, the environmental fate of many of the chemicals we use is utterly unknown. This includes a majority of the hundreds of non-metal-containing organic pigments and dyes used in our materials. Since no one knows what happens to them in the waste stream, they cannot be regulated and may be labeled "biodegradable." Experts suspect that many dyes and pigments have long-term adverse effects on people and perhaps on the environment. Studies are underway.

> ***Biodegradable* really means:** *You won't get into trouble if you flush or trash the product, but there are no guarantees related to your health while you use it. And the product may not be safe for the environment in the long run either.*

Water-Based

Many people assume that water-based products are safe. But the material safety data sheets from the manufacturers reveal that many water-based products contain both water and solvents. Examples of water-based products that often contain solvents include spray cleaners, paint strippers, felt-tip markers, and latex wall paints.

The solvents most often found in water-based products are glycol ethers. These solvents can usually be absorbed through your skin and can penetrate several types of heavy chemical gloves without changing the glove's appearance. Some glycol ethers cause reproductive problems in humans, and birth defects and atrophy of the testicles in experimental animals.

> ***Water-based*** **really means:** *Water is probably an ingredient, but the product may also contain solvents and other toxic substances. Consult the material safety data sheet for further information.*

Contains No VOCs

Strictly translated, VOC means "volatile organic chemical." But on labels, VOC refers to those volatile solvents that are regulated because they create smog. The technical definition of a VOC is any chemical that "participates in atmospheric photochemical reactions" (40 CFR Part 51, Section 100). However, many solvents, such as acetone and ethyl acetate, react negligibly in the atmosphere. These are called "exempt compounds" and are not labeled as VOCs.

> ***Contains no VOCs*** **really means:** *There are no solvents in this product that will react in the atmosphere to create smog, but the product can be full of exempted solvents that are toxic to users.*

GRAS (Generally Recognized As Safe)

Use of the term "GRAS" on labels or in the product's literature can mislead you into thinking the product is safe enough to eat. GRAS is an FDA term that means many different things. Some GRAS substances are allowed to be added to food in specified amounts, and others are restricted to use in food packaging, as disinfectants and soaps to sanitize food processing utensils, and similar uses in which they would not be ingested. For example, highly hazardous substances can be GRAS for use in food packaging if manufacturers can prove they do not migrate from the package into food.

Even substances that are allowed in food in very small amounts may not be safe if internalized through other routes of entry into the body. For example, fumed silica is allowed in some foods. It is not hazardous to eat, but it is hazardous to inhale.

Some substances are classified as GRAS simply because they have been used for years and no one has bothered to test them. An example was seen in the news recently. Phenolphthalein was GRAS for use in over-the-counter laxatives for about ninety years. It was finally tested and found to cause cancer in a large percentage of test animals. The FDA has withdrawn its GRAS status.

> ***GRAS* really means:** *Either the ingredient has been used for years without testing or the FDA has approved it for some very limited use related to food, food packaging, etc. It should not be assumed safe to ingest, and GRAS status certainly is no guarantee of safety by inhalation.*

Nontoxic

Under the United States Federal Hazardous Substances Act and the Canadian Federal Hazardous Products Act, toxic warnings are required on products capable of causing acute (sudden onset) hazards. Products requiring warning labels are identified by tests that expose animals to a single dose or period of exposure by skin or eye contact, inhalation, and ingestion. These tests are incapable of identifying products that cause long-term hazards such as cancer, birth defects, allergies, and chronic illnesses.

To illustrate the inadequacy of this labeling, powdered asbestos could be labeled "nontoxic" on the basis of these acute tests. All the animals exposed to asbestos will appear perfectly healthy after two weeks, because cancer and asbestosis take much longer to develop.

> ***Nontoxic* on a consumer product really means:** *The ingredients don't kill half or more of the animals in short-term toxicity tests, but there are no long-term guarantees.*

Nontoxic Art Materials

In the past, asbestos-containing products such as instant papier-mâchés and clays were routinely labeled "nontoxic." This should not happen today, because a new law, the Labeling of Hazardous Art Materials Act (LHAMA), requires warning labels on products containing known, chronically hazardous substances. However, LHAMA has serious deficiencies:

- Substances requiring labeling must be *known* to possess chronic hazards. Yet, most substances used in art materials, especially organic colorants, have never been tested for chronic hazards. They can be labeled "nontoxic" by default! Many "nontoxic" pigments and dyes are members of classes of chemicals that clearly are carcinogens.

- Some tests used to determine toxicity are faulty. For example, a test is sometimes used in which materials are placed in acid to determine if the toxic substances in the product will be released in the digestive tract. This test does not consider that in addition to acid and water, digestive tracts use bases, enzymes, cellular activities, heat, and movement to dissolve materials. Only after people got lead poisoning from "nontoxic" ceramic glazes were these tests abandoned by glaze manufacturers. But many paint manufacturers still use the acid tests!
- Faulty methods are used to determine when the amounts of toxic substances in art materials will result in a significant exposure. Toxicologists who estimate exposures often do not take fully into account the artist's penchant for intimate contact with materials, crowded workspaces or tiny home studios, poor ventilation, lack of sinks, and other conditions.
- The hazards of materials used in ways other than the label directs are not considered. Theater artists and teachers traditionally use materials "creatively." For example, melting crayons for candlemaking, batik resist, or for other processes causes these "nontoxic" products to release highly toxic gases and fumes from decomposing wax and pigments.
- Enforcement of the laws is so weak that lots of nonconforming imported products are on the market. For example, in 1995, a cameraman and reporter from Channel 9 in New York went with me to a major art material outlet. That night on the evening news, we showed viewers about a dozen imported art products that are in violation of our hazardous material labeling standards that were being sold illegally.

***Nontoxic* labels on art materials really mean:** *A toxicologist who has not taught or created with the materials says that users will not be exposed to significant amounts of any* known *chronically toxic ingredients. Untested ingredients may be labeled "nontoxic" even though they are expected to be toxic or cancer-causing on the basis of their chemical structure. Use other than directed at your own risk. Some imported products don't conform at all.*

Natural

There is nothing inherently safe in substances derived from nature. This is obvious if we just think for a moment about turpentine, wood dust, molds, poison ivy, cocaine, jimsonweed, curare, hemlock, tobacco, and so on. In fact, natural toxins like ricin, from the castor bean plant, or botulism toxin are thousands of times more toxic than human-made chemicals like sodium cyanide or the methyl isocyanate that killed over two thousand people in Bhopal, India! Judge natural products just as you would synthetic ones. For example, take a cold, hard look at just one commonly used natural product from citrus fruit.

Citrus Oil: A Case in Point. Citrus oil and its major component, d-limonene, are derived from the rinds of citrus fruit. Products containing d-limonene are touted as "natural" and "biodegradable." Many paint thinners and strippers, cleaning agents, and hand cleaners contain citrus solvents.

Advertisements for these products often emphasize that the FDA allows small amounts in food as an additive. The advertisements fail to mention that d-limonene is one of Mother Nature's own pesticides. She put it in the rinds to protect her fruit from insects. It kills flies efficiently enough to be registered with the EPA as both an "active ingredient" and an "inert ingredient" in commercial pesticides.

D-limonene and citrus oil can also be contaminated with other pesticides from the spraying of the fruit in the orchards. The EPA once proposed revoking the use of citrus oil in food, because it usually is contaminated with cancer-causing pesticides.

***Natural* really means:** *The manufacturer wants you to prejudge the product's toxicity on the basis of its origins. Instead, ask questions and look up the hazards. Highly toxic products are manufactured by both God and Goodyear.*

Use with Adequate Ventilation

Many people think ventilation means keeping a window or door open while using the product. Actually, this label only indicates that the product contains something toxic that becomes airborne during the product's use. The ventilation required must be sufficient to keep the airborne substance below levels considered acceptable for industrial air quality. Sufficient ventilation could vary from a simple exhaust fan to a specially designed local exhaust system, depending on the amount of the material and how it is used.

In order to plan such ventilation, you must know exactly what substance the product gives off and at what rate. Ironically, this is often precisely the information the manufacturer excludes from the label. In order to plan ventilation, first get the material safety data sheet from the manufacturer.

***Adequate ventilation* really means:** *Something toxic gets airborne during use. You need to know what it is, how toxic it is, and how much ventilation you will need for the amount of the substance you use. You should refer to a material safety data sheet, and you may need professional advice.*

Industrial or Professional Use Only

Products carrying this label are not supposed to be readily available to general consumers and should never be used by children or untrained adults. Rules for the types of information and warning symbols that conform to the right-to-know law can be obtained from your local department of labor. This label warns workers that they should be skilled in the use of the product and should have a material safety data sheet as a guide to safe use of the product.

6 Material Safety Data Sheets

WHAT ARE THEY?

Material Safety Data Sheets (MSDSs) are forms that provide information on a product's hazards and the precautions required for its safe use in the working environment. MSDSs are usually filled out by the product's maker. The quality of the information varies, depending on the diligence of the individual manufacturer. However, manufacturers are responsible to their respective government agencies for the accuracy of information they provide. MSDSs are essential starting points for collection of information, but they should not be considered complete sources of information on their own.

WHERE CAN I GET THEM?

MSDSs can be obtained by writing or calling manufacturers, distributors, or importers of the product. Employers can require MSDSs as a condition of purchase. If makers or suppliers do not respond to requests for MSDSs, send copies of your requests and other pertinent information to the agency responsible for enforcing your federal or local hazard communication law. These agencies usually get very good results.

Large Internet data banks often are not a good source of MSDSs, unless the Web site is run by the actual manufacturer of your product. Some of the large general MSDS databases are not updated frequently, and you will be downloading old material. This is especially important today, when many product formulations are changing frequently due to the phasing out of many chemicals that damage the environment.

WHO NEEDS THEM?

The U.S. OSHA Hazard Communications Standard and the similar Canadian standard (WHMIS) regulations require employers to make MSDSs available to all those who use or could be exposed to potentially hazardous products in the workplace. In addition, employers and administrators are responsible for training employees to understand and use MSDS information.

WHERE SHOULD THEY BE KEPT?

MSDSs are required to be readily available to workers during all working hours. It is a good idea to have a central file for all MSDSs, and then to file or display copies where products are actually used and stored.

WHAT DO THEY LOOK LIKE?

To begin with, the words "Material Safety Data Sheet" must be at the top. Some manufacturers resist sending MSDSs, sending instead sheets labeled "Product Data Sheet," "Hazard Data Sheet," or other improper titles. These information sheets do not satisfy the laws' requirements.

Although MSDSs must contain the same basic information, the actual sheets may look very different from each other. The order in which information is presented may differ; some are computer-generated, and some are filled out on OSHA forms. Figure 6.1 is a copy of the United States Material Safety Data Sheet form (the OSHA 174 form).

REQUIRED DATA

All of the information in table 6.1 should be on the form. In general, MSDSs should identify the product chemically (with the exception of trade secret products), identify any toxic ingredients, list its physical properties, its fire and explosion data, its acute and chronic health hazards, its reactivity (conditions that could cause the product to react or decompose dangerously), tell how to clean up large spills and dispose of the material, identify the types of first aid and protective equipment needed to use it safely, and discuss any special hazards the product might have.

Material Safety Data Sheet
May be used to comply with
OSHA's Hazard Communication Standard,
29 CFR 1910.1200. Standard must be
consulted for specific requirements.

U.S. Department of Labor
Occupational Safety and Health Administration
(Non-Mandatory Form)
Form Approved
OMB No. 1218-0072

IDENTITY *(As Used on Label and List)*	*Note: Blank spaces are not permitted. If any item is not applicable, or no information is available, the space must be marked to indicate that.*

Section I

Manufacturer's Name	Emergency Telephone Number
Address *(Number, Street, City, State, and ZIP Code)*	Telephone Number for Information
	Date Prepared
	Signature of Preparer *(optional)*

Section II — Hazardous Ingredients/Identity Information

Hazardous Components (Specific Chemical Identity; Common Name(s))	OSHA PEL	ACGIH TLV	Other Limits Recommended	% *(optional)*

Section III — Physical/Chemical Characteristics

Boiling Point		Specific Gravity (H_2O = 1)	
Vapor Pressure (mm Hg.)		Melting Point	
Vapor Density (AIR = 1)		Evaporation Rate (Butyl Acetate = 1)	
Solubility in Water			
Appearance and Odor			

Section IV — Fire and Explosion Hazard Data

Flash Point (Method Used)	Flammable Limits	LEL	UEL
Extinguishing Media			
Special Fire Fighting Procedures			
Unusual Fire and Explosion Hazards			

(Reproduce locally) OSHA 174, Sept. 1985

FIGURE 6.1. MATERIAL SAFETY DATA SHEET (U.S.)

Section V — Reactivity Data

Stability	Unstable		Conditions to Avoid
	Stable		

Incompatibility (*Materials to Avoid*)

Hazardous Decomposition or Byproducts

Hazardous Polymerization	May Occur		Conditions to Avoid
	Will Not Occur		

Section VI — Health Hazard Data

Route(s) of Entry:	Inhalation?	Skin?	Ingestion?

Health Hazards (*Acute and Chronic*)

Carcinogenicity:	NTP?	IARC Monographs?	OSHA Regulated?

Signs and Symptoms of Exposure

Medical Conditions
Generally Aggravated by Exposure

Emergency and First Aid Procedures

Section VII — Precautions for Safe Handling and Use

Steps to Be Taken in Case Material Is Released or Spilled

Waste Disposal Method

Precautions to Be Taken in Handling and Storing

Other Precautions

Section VIII — Control Measures

Respiratory Protection (*Specify Type*)

Ventilation	Local Exhaust	Special
	Mechanical (*General*)	Other

Protective Gloves	Eye Protection

Other Protective Clothing or Equipment

Work/Hygienic Practices

Page 2

★ U.S.G.P.O. 1986-491-529/45775

FIGURE 6.1. MATERIAL SAFETY DATA SHEET (U.S.) (CONTINUED)

DATA REQUIRED ON MATERIAL SAFETY DATA SHEETS (MSDSs)

MSDS forms are organized differently by different manufacturers. But the same data must be present somewhere on the form. According to 29 CFR 1910.1200(g)(2)(i)-(xii) and OSHA form 174, the required data includes:

(i) *Identity of products.* Both chemical and common names for single substances or mixtures that have been studied for hazards as a whole. For all other products, MSDSs must supply chemical and common name(s) for all hazardous ingredients comprising 1.0 percent of the product or carcinogens comprising 0.1 percent (or less, if smaller amounts are hazardous).

(ii) *Physical and chemical characteristics.* Boiling point, vapor pressure, vapor density, solubility in water, specific gravity, melting point, evaporation rate.

(iii) *Physical hazards.* Potential for fire, explosion, and reactivity, flash point, flammable limits, stability, incompatibility, hazardous decomposition products, hazardous polymerization.

(iv) *Health hazards.* Both acute and chronic, signs and symptoms of exposure, and medical conditions aggravated by exposure.

(v) *Primary route(s) of entry.* Skin, inhalation, and ingestion.

(vi) *Exposure limits.* OSHA PELs, ACGIH TLVs, and any other exposure limit used or recommended by the chemical manufacturer, importer, or employer.

(vii) *Carcinogenicity.* Whether the chemical is listed as a carcinogen by NTP, IARC, or OSHA.

(viii) *Precautions for safe handling and use.* Hygienic practices, protective measures during repair and maintenance, and procedures for cleanup of spills and leaks.

(ix) *Control measures.* Ventilation, respiratory protection, gloves, eye protection, protective clothing or equipment, hygienic practices.

(x) *Emergency and first aid procedures and firefighting methods and hazards.*

(xi) *Date of preparation or last revision.*

(xii) *Name, address, and telephone number of responsible party.*

HOW TO READ MSDSs

It would take a small book to explain how to read an MSDS and define all the terms you are likely to encounter. Fortunately, there already are many small books and pamphlets on this subject. Your local department of labor, the American Lung Association, and many other organizations publish such items (see Bibliography). You can also obtain a free copy of a data sheet called "Understanding the MSDS" from ACTS by sending a self-addressed, stamped envelope with your request to:

Arts, Crafts and Theater Safety (ACTS)
181 Thompson Street, # 23
New York, New York 10012-2586,
(212) 777-0062.

However, there are certain pieces of information that are particularly important on MSDSs, and they are covered below.

The Responsible Party

The Material Safety Data Sheet must have the manufacturer's name, address, and emergency telephone number. The name and address on the MSDS must match the name and address on the product's label. If you notice that the two names are different, it is likely that one company repackaged another manufacturer's products under its own label, but sent out the MSDS from the original manufacturer. This is a violation of the law, and the MSDS must bear the repackager's name as the responsible party.

Some importers violate this law. Importers' names and addresses must be on the MSDS, because they are considered the "manufacturer of record." They are not legally allowed to send out a foreign MSDS. Instead, they must produce an MSDS that meets U.S. requirements, in English, and lists a twenty-four-hour emergency number in this country.

Date of Preparation

The date on which the document was written or last revised must appear on the MSDS. In Canada, three-year-old MSDSs are automatically invalid. In the United States, MSDSs should be reviewed frequently, and those that fail to incorporate up-to-date information are easy to get updated. Send a letter to the company pointing out omissions, and include a reminder that OSHA's Hazard Communication Standard, 29 CFR 1910.1200(g)(6), gives the manufacturer only three months to incorporate significant, new hazard information. Send a carbon copy to OSHA!

Identity of Ingredients

Only the chemical and common names of the ingredients must be listed under the federal law. However, good MSDSs provide the chemical class, other synonyms, and the Chemical Abstract Service registration numbers.

Chemical Abstract Service (CAS) Registration Numbers

Many state OSHA laws require CAS identification on their MSDSs. CAS registration numbers are assigned sequentially and have no chemical significance. However, they prevent confusion between chemicals with very similar names. For example:

SIMILAR NAMES	CAS RN	TOXICITY AND USE
benzene	71-43-2	an extremely toxic solvent
benzine	64475-85-0	a low toxicity petroleum solvent
benzidine	92-87-5	an extremely toxic dye intermediate

Trade Secrets and Proprietary Ingredients

If the manufacturer can prove that certain ingredients provide an advantage over competitors who do not know of them or use them, the manufacturer can register these ingredients as trade secrets. Once registered, only the words "trade secret" or "proprietary" and the registration number are required on this section of the MSDS. Make sure there is a registration number; if none is listed, the ingredient is not a legitimate trade secret.

Less legitimate methods may be used by some companies to obscure the identity of ingredients. For example, they might omit the name "acetone" and identify the chemical only by the rarely used synonym, "dimethylformaldehyde."

OSHA requires that trade-secret information be disclosed to health professionals if they suspect their patients are affected adversely by the product. However, I have found that this transaction rarely occurs, because those who receive proprietary information must sign a confidentiality agreement. This agreement carries severe financial penalties if the information is revealed, even accidentally, to anyone, including the patient. I find that doctors and other health professionals are reluctant to sign these binding agreements.

Air Quality Limits

One of the most difficult and most useful pieces of data on the MSDS are the workplace air quality limits. There are two that must be reported: the OSHA-permissible exposure limits (PEL) and the American Conference of Governmental Industrial Hygienists (ACGIH) Threshold Limit Value (TLV). Good MSDSs will also list any other exposure limit used or recommended by the chemical manufacturer or importer.

TABLE 6.2 EXAMPLES OF THRESHOLD LIMIT VALUES (TLVs) IN PARTS PER MILLION (ppm)

SUBSTANCE	TLV
carbon dioxide	5,000 ppm
ethanol, some spray can propellants	1,000 ppm
xylene, turpentine, mineral spirits	100 ppm
n-hexane, toluene	50 ppm

The TLVs are usually safer limits than the PELs. This is because OSHA has difficulty in updating their PELs over objections from industry. The TLVs are useful for selecting safer chemicals. The higher the chemical's TLV, the more is allowed in the workplace air, and the more can be inhaled before a healthy adult can expect adverse effects. To use TLVs properly, see the more detailed explanation of the TLV in chapter 8.

Evaporation Rate

The more slowly the chemical evaporates, the less vapor it creates, and the less hazardous it is to work with. In fact, some very toxic solvents that evaporate slowly may not be as hazardous to use as less toxic ones that evaporate quickly.

Acute and Chronic Effects

Short-term animal tests provide acute data on many chemicals. Chronic data, however, take years to develop. Most chemicals have never been studied for long-term effects. The lack of chronic data should never be interpreted as an indication of safety.

Carcinogenicity

The cancer ratings of three agencies must be reported on MSDSs: the National Toxicology Program's (NTP) *Annual Report on Carcinogens* (latest edition), the International Agency for Research on Cancer's (IARC) *Monographs* (latest edition), and OSHA. However, the majority of the substances we use have never been studied for cancer effects at all. When an MSDS reports that none of the three agencies consider a substance a carcinogen, it is most likely that none of the agencies have even reviewed the substance, because there is not enough data to judge.

Releases and Spills

Manufacturers' directions for dealing with spills and releases are often vague or designed for factory or chemical laboratory use. People dealing with spills need special training, because no MSDS can provide spill and release information to cover all contingencies. Procedures must be adapted to accommodate the size of the spill, the type of surface it spills on, whether the spill can get into drains in violation of EPA rules, whether the ventilation will spread toxic vapors into other areas, whether there are reactive chemicals nearby, and much more.

Control Measures

Some MSDSs provide extremely overcautious advice about respiratory protection, gloves, and eye protection. I have seen MSDSs that advise wearing a full bodysuit and a self-contained breathing apparatus (SCBA) for chemicals intended for use on a Q-Tip™. Actually, those were reasonable precautions when the primary manufacturer wrote them for workers wading in vats of the stuff. But when such precautions are on MSDSs for consumer products, they indicate that the secondary supplier was too ignorant or too unconcerned about our safety to tailor the control measures to the label-directed use.

On the other hand, some MSDSs provide too few precautions and rely on useless phrases such as "wear protective goggles," "avoid inhalation," and "use respiratory protection." Good MSDSs are more specific about the types of protective equipment needed.

These are just some of the ways MSDS data can be used. Clearly, MSDSs do not provide perfect information. But even seriously flawed MSDSs can be useful to document a company's lack of expertise and/or regulatory compliance. And they are required reading in any workplace.

7

Plans, Programs, and Personal Protection

There are a number of programs that must be in place before work occurs on any site. Included are programs that provide for emergency egress and evacuation, fire prevention and protection, medical response, and first aid. There must be additional programs in place if workers need to wear protective gear, such as gloves or hard hats.

EMERGENCY PLANS

The OSHA regulations that apply to egress, emergency evacuation, and fire prevention are taken directly from the National Fire Protection Association's (NFPA) Life Safety Code. These NFPA standards are also used as the source of local fire codes and are enforced by fire marshals as well.

The emergency egress and evacuation regulations (1910.35-38, 1926.150) require employers to formally train to know what the alarms or other warnings mean, where to exit, where to meet for a head count, and so on. Theater workers are probably most familiar with the exit sign requirements (1910.37). A written plan is required in 1910.38.

Fire prevention regulations (1910.155-166, 1926.150) cover all types of fire protection systems, fire brigades, and training. The rules on use of fire extinguishers in permanent venues and construction (location) sites are in 1910.157 and 1926.150, respectively. These regulations require workers to have formal training in their use when they are first employed and at least once a year thereafter.

MEDICAL RESPONSE

A program for dealing with medical emergencies must be in place in any workplace (1910.151, 1926.50). This often means consulting with local authorities and hospitals to get estimates of their response time and services. This is especially important on temporary locations, which come under the construction standard.

The applicable rules in 1926.50, medical services and first aid, subpart D, require that:

1. The employer shall ensure the availability of medical personnel for advice and consultation on matters of occupational health
2. Provisions shall be made prior to commencement of the project for prompt medical attention in case of serious injury
3. In the absence of an infirmary, clinic, hospital, or physician that is reasonably accessible in terms of time and distance to the work site, which is available for the treatment of injured employees, a person who has a valid certificate in first aid training from the U.S. Bureau of Mines, the American Red Cross, or equivalent training that can be verified by documentary evidence, shall be available at the work site to render first aid

The employer has a number of options in meeting these requirements, but however this is done, each employee should be repeatedly trained to know precisely what their individual responses should be to medical emergencies. Emergency telephone numbers usually are posted near each telephone, and a chain of command is established for rapid response within the workforce.

FIRST AID

Medical first aid equipment and supplies also must be on site, and people must know how to use them.

Eyewash Stations

Emergency eyewashes and showers are needed if corrosive chemicals are used in the venue. This would include eye-damaging chemicals such as paint solvents, curing agents for polyester and urethane resins, lye, many boiler additives and air-conditioning chemicals, acetic acid for photochemical stop baths, and the like.

The squeeze bottle containers of eyewash fluid are not approved for this purpose. The eyewash stations must instead meet the OSHA requirements, which reference an American National Standards Institute standard (ANSI Z 87.1). The standard dictates location and design features of eyewash standards and emergency showers. They should be run once a week for three minutes to ensure they are working and to clear the plumbing of microorganisms. First aid kits should be equipped with any special items needed.

First Aid Kits

First aid kit requirements are in 1910.151(b) and 1926.50. The supplies used to require the approval of a consulting physician, but now OSHA provides a list of basic first aid supplies in a new, *non-mandatory* appendix A to sec. 1910.151. This appendix references the American National Standards Institute (ANSI A308.1-1978, "Minimum Requirements for Workplace First-Aid Kits"). In the future, appendix A may reference the new ANSI Z308.1-1998.

This new ANSI standard gives the employer more flexibility in determining the first aid supplies needed for workplace hazards, and unit packages are now color-coded as follows:

Blue—Antiseptics
Yellow—Bandages
Red—Burn Treatment
Orange—Personal Protective Equipment
Green—Miscellaneous

Bloodborne Pathogens Training

Any worker expected to deal with even minor cuts must have training on how to deal with body fluids, under the bloodborne pathogens standard (1910.1030). This regulation applies to most shops, including the costume shop, where needle sticks are a common occurrence. The rule requires that sharp items that are contaminated with blood or other body fluids be disposed of in a medical-biological-hazards (sharps) container, that workers be trained to know about diseases transmitted by body fluids, how to protect themselves with surgical gloves, and how to remove gloves without contaminating their hands. (See also section on gloves below.)

PERSONAL PROTECTIVE EQUIPMENT

OSHA now requires employers to have formal written programs and training if their workers wear protective equipment such as gloves, eyewear, aprons, hard hats, and the like. This training is sorely needed in our business. For example, I have observed scenic artists using small power tools who were wearing eye protection rated for chemical splash protection, when impact resistance was needed. And I have seen them wearing gloves that are permeable to the very chemicals they were using.

The Regulations

Under the Personal Protective Equipment (PPE) standard (1926.28,1910.132-133), employers must first assess the hazards in the workplace. They can either hire someone to do this assessment or do it themselves, if they have the experience. The best person to do it is someone who is either a certified industrial hygienist or certified safety professional.

Assessment is done by following the workers on their jobs and identifying the points at which protective equipment is needed. For example, the assessor might note that steel-toed shoes are needed when counterweights are loaded, or eye protection is needed when workers use woodworking equipment. All the hazards of the various jobs must be compiled in writing. The employer must certify that the workplace has been assessed.

Next, employers or their consultants must select the appropriate equipment to protect the employees, communicate that decision to the employees, and make sure that the equipment fits each employee. They also must formally train the workers to know when they need to wear the equipment; what type is necessary; how to don, doff, adjust, and wear the equipment; the limitations of the equipment; and the proper care of the equipment.

Then, the workers must be trained. At the end of training, the trainer must give workers a quiz or find some other way to ensure that they comprehended the training. Training records and quizzes must be kept and retraining done whenever new hazards or equipment are used, new personnel are hired, or when workers need refresher training.

Gloves

A good example of why training is needed can be seen in something as simple as wearing gloves. Figure 7.1 provides information on the different types of gloves that may be worn for various purposes. Figure 7.2 provides an outline of information that should be understood by all users of chemical resistance gloves. It has been my experience that many theater workers do not know that chemicals can permeate their gloves without altering their appearance.

Face and Eye Protection

Suitable face and eye protection in the form of goggles or shields will guard against a variety of hazards, including impact (from chipping, grinding, etc.), radiation (welding, carbon arcs, lasers, etc.), and chemical splash (solvents, acids, etc.). Such protective equipment should be used only if it meets the standards of the American National Standards Institute's *Practice for Occupational and Educational Eye and Face Protection* (ANSI Publication Z87.1). Choose eye protection appropriate to the hazard (see figures 7.3 and 7.4).

Hearing Protection

Ear-damaging levels of sound can be produced by the machinery in many shops, by sound equipment, pyrotechnic blasts, and musical instruments. It has long been known, for example, that orchestral musicians sitting in front of horns or

TYPES OF WORK GLOVES

Disposable or surgical gloves. Only for use against mild irritants and bloodborne pathogens, never for chemical protection.

Fabric gloves. Made of cotton or fabric blends, they improve grip and protect against mild heat and cold and mildly abrasive materials.

Leather gloves. Use against sparks (welding), for scraping rough surfaces, and brief handling of hot materials. Can be used in combination with an insulated liner when working with electricity.

Metal mesh. Used to protect wearers from cuts and scratches from cutting tools, sharp instruments, sorting/cutting glass, etc.

Aluminized fabric gloves. Designed to insulate against high heat, such as molten metal, old-fashioned carbon arc spotlights, etc.

Chemical resistance gloves. Usually made of natural rubber; synthetic polymers such as butyl, butadiene, neoprene, acrylonitrile, or polyvinyl alcohol; or vinyl rubbers or plastics. It is necessary to consult the manufacturer's "glove chart" for recommendations on which type will protect users from specific oils, solvents, corrosives, etc.

Natural rubber gloves. Special training is needed for anyone using gloves containing natural rubber, since allergies to these products can be life-threatening. See the chapter on plastics for a complete description of the hazards and precautions to take with rubber latex gloves and rubber molding products.

FIGURE 7.1

near timpani show hearing loss in time. And recently, noise levels at a coliseum were assessed during a monster truck show and found to be at levels requiring hearing protection.[1]

[1]NIOSH, HETA 98-0093-2717, U.S. Hot Rod Monster Truck and Motocross Show, November 1998.

CHEMICAL RESISTANCE GLOVES: TRAINING OUTLINE (29 CFR 1910.132-3)

LOGISTICS AND GENERAL USE

- Location of supply, how to get and replace gloves at your work site.
- How to check for holes and damage before putting them on.
- How to take them off so the contaminants don't get on your skin.
- How to clean or discard them.

HOW TO PREVENT SKIN PROBLEMS

- Using the right glove for the purpose.
- Never sharing gloves unless they are cleaned and disinfected.
- Washing hands and applying emollient after removing gloves.
- Watching for skin reactions to gloves, chemicals, sweat, etc.

TECHNICAL INFORMATION

- Gloves are made of several types of plastics and natural rubber.
  —Plastics (e.g., neoprene, nitrile, PVC, butyl)
    - Each type resists different chemicals.
    - Allergies to them are rare, usually caused by additives.

  —Natural rubber
    - Allergies common: rashes, asthma, anaphylactic shock.
    - The word "hypoallergenic" is prohibited by the FDA on rubber gloves, since they cannot be made safe.
    - Powdering on inside of gloves can make allergies worse.

Surgical gloves, whether rubber or plastic, are never for chemical use. They only protect the hands from mild, waterborne irritants, like soap and water, and body-fluid pathogens.

- Charts listing degradation/penetration for specific gloves
  Glove charts are available from manufacturers. Use only the chart from the manufacturer of your particular gloves. Tell workers where the glove chart is located at the work site, and teach them how to use it.

- Degradation of glove materials
  Degradation is caused by chemical exposure, direct sunlight, and age. It can usually be seen visually when the gloves get "gummy," cracked, or brittle. Such gloves should be immediately discarded.

- Penetration of gloves by chemicals
  Chemicals can penetrate gloves and expose the user's skin to the chemical. Penetration creates no visible change in the gloves' appearance. This is why it is necessary to use material safety data sheets to identify the chemicals in the products, and then use the glove chart to choose gloves that are capable of resisting those specific chemicals. Never use Vaseline or barrier creams on your skin when wearing gloves, since these materials worsen penetration.

Demonstrate how the chart was used to select the glove the worker will be using.

FIGURE 7.2

EYE PROTECTION: SELECTION CHART

AMERICAN NATIONAL STANDARD Z87.1-1989

SELECTION CHART

		ASSESSMENT SEE NOTE (1)	PROTECTOR TYPE	PROTECTORS	LIMITATIONS	NOT RECOMMENDED
IMPACT	Chipping, grinding, machining, masonry work, riveting, and sanding.	Flying fragments, objects, large chips, particles, sand, dirt, etc.	B,C,D, E,F,G, H,I,J, K,L,N	Spectacles, goggles faceshields SEE NOTES (1) (3) (5) (6) (10) For severe exposure add N	Protective devices do not provide unlimited protection. SEE NOTE (7)	Protectors that do not provide protection from side exposure. SEE NOTE (10) Filter or tinted lenses that restrict light transmittance, unless it is determined that a glare hazard exists. Refer to OPTICAL RADIATION.
HEAT	Furnace operations, pouring, casting, hot dipping, gas cutting, and welding.	Hot sparks	B,C,D, E,F,G, H,I,J, K,L,*N	Faceshields, goggles, spectacles *For severe exposure add N SEE NOTE (2) (3)	Spectacles, cup and cover type goggles do not provide unlimited facial protection. SEE NOTE (2)	Protectors that do not provide protection from side exposure.
		Splash from molten metals	*N	*Faceshields worn over goggles H,K SEE NOTE (2) (3)		
		High temperature exposure	N	Screen faceshields, Reflective faceshields. SEE NOTE (2) (3)	SEE NOTE (3)	
CHEMICAL	Acid and chemicals handling, degreasing, plating	Splash	G,H,K *N	Goggles, eyecup and cover types. *For severe exposure, add N	Ventilation should be adequate but well protected from splash entry	Spectacles, welding helmets, handshields
		Irritating mists	G	Special purpose goggles	SEE NOTE (3)	

FIGURE 7.3. ANSI PROTECTION STANDARDS

DUST	Woodworking, buffing, general dusty conditions.	Nuisance dust	G,H,K	Goggles, eyecup and cover types	Atmospheric conditions and the restricted ventilation of the protector can cause lenses to fog. Frequent cleaning may be required.	
OPTICAL RADIATION	WELDING: Electric Arc		O,P,Q	TYPICAL FILTER LENS SHADE / PROTECTORS — SEE NOTE (9) 10-14 Welding Helmets or Welding Shields	Protection from optical radiation is directly related to filter lens density. SEE NOTE (4). Select the darkest shade that allows adequate task performance.	Protectors that do not provide protection from optical radiation. SEE NOTE (4)
	WELDING: Gas		J,K,L, M,N,O, P,Q	SEE NOTE (9) 4-8 Welding Goggles or Welding Faceshield	SEE NOTE (3)	
	CUTTING			3-6		
	TORCH BRAZING			3-4		
	TORCH SOLDERING		B,C,D, E,F,N	1.5-3 Spectacles or Welding Faceshield		
	GLARE		A,B	Spectacle SEE NOTE (9) (10)	Shaded or Special Purpose lenses, as suitable. SEE NOTE (8)	
16						

FIGURE 7.3. ANSI PROTECTION STANDARDS (CONTINUED)

EYE PROTECTION PRODUCTS

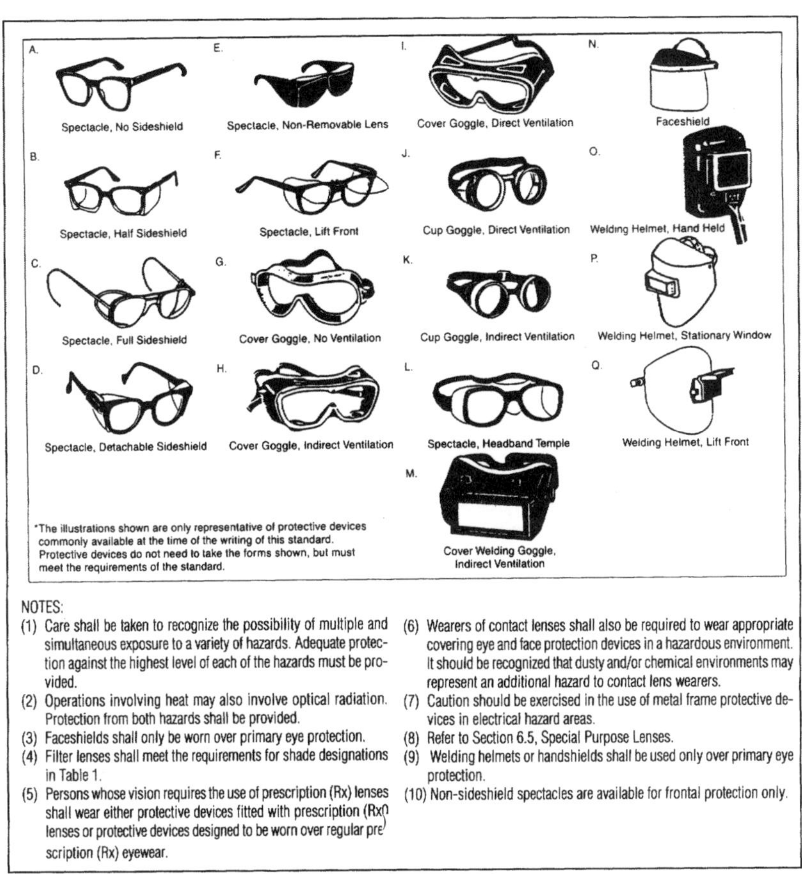

FIGURE 7.4. EYE PROTECTION PRODUCTS

OSHA has set limits for noise for all workplaces, but the limits are not set low enough to protect everyone's hearing, they are often hard to enforce, and it takes special equipment to measure and document exposure adequately. However, the OSHA regulations require employers to assess noise levels (29 CFR 1910.95 or 1926.52). If the noise levels are at or near permissible noise levels, employers must provide hearing protection, audiograms, and formal training about noise and the use of protection devices.

In general, if you need to raise your voice to be heard by someone two feet from you, you probably need hearing protection. These levels are usually in the range of eighty-five decibels or more (see figure 7.5). Workers or students who are hearing-impaired should be evaluated medically to determine if a noisy environment can damage their hearing further.

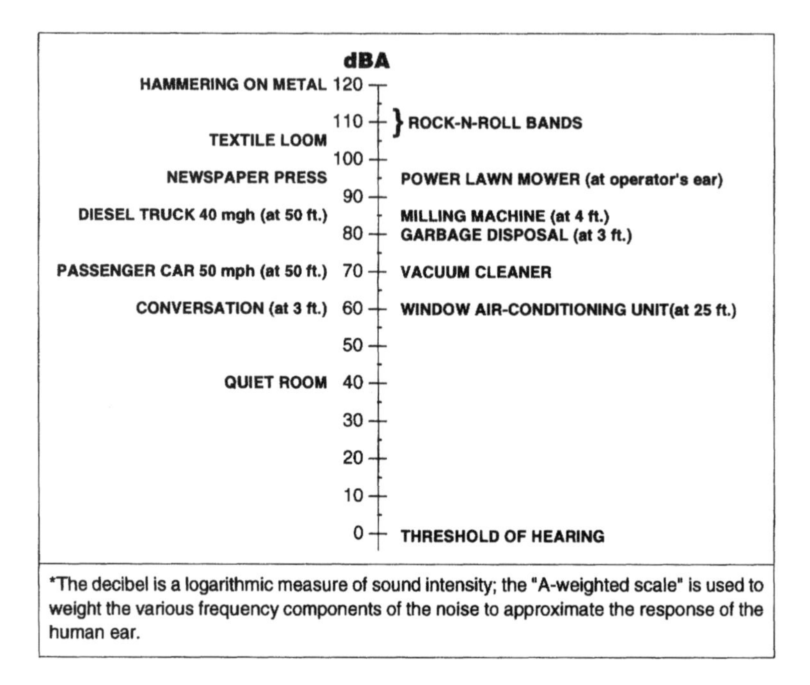

FIGURE 7.5. TYPICAL A-WEIGHTED NOISE LEVELS IN DECIBELS (DBA)*

In shops, the best way to protect your hearing is to eliminate or reduce noise. Machinery vibration can be dampened by installing rubber gaskets or putting rubber pads under it. Keeping machinery in good repair and oiled also helps, because properly maintained machines run more quietly. When making new purchases, choose quieter brands of machinery.

If these methods do not lower noise levels sufficiently, then you should wear hearing protection, such as earplugs and muffs. A good selection of these appears in many safety catalogs. Under OSHA regulations, the type of hearing protection is matched to the character of the noise in the workplace.

Other Protective Equipment

Hard hats, various types of footwear such as steel-toed shoes, aprons and various types of protective clothing, and more may be needed. There are standards for selection and use for each of these types of protective gear that must be followed.

Protective Equipment Rules

- Select qualified personnel to set up the OSHA PPE program and provide the training. The resumes of these people also belong in the program file.
- Get expert advice from more than one source when choosing protective equipment, purchasing quieter machinery, etc. Never rely solely on equipment manufacturers or distributors for information.
- Collect catalogs and other good sources for types of protective equipment (see Bibliography).
- Determine if any employees have special problems regarding protective equipment. For example, people with beards should not wear air-purifying respirators, and people with dermatitis may need special hand protection.
- Enforce proper use of protective equipment. Ventilation systems are useless if they are not turned on, and hearing is not protected by unused earmuffs.
- Develop a schedule for regular inspection, repair, and replacement of protective equipment.

8 The Air We Breathe

The air around us isn't all air! Many hazardous contaminants can be present in the air in the form of gases, vapors, mists, fumes, and dusts. The physical nature of these contaminants must be understood in order to know how hazardous they are and how to protect ourselves from them.

AIR CONTAMINANTS

Gases

Scientists define gas as "a formless fluid" that can "expand to fill the space that contains it." We can picture this fluid as many molecules moving rapidly and randomly in space.

Air, for example, is a mixture of different gases—that is, different kinds of molecules. Even though each different gas has a different molecular weight, the heavier gases will never settle out, because the rapid movement of the molecules will cause them to remain mixed forever. Gases created on stage or in shops will also diffuse evenly through the air, in whatever space is available to them.

There are exceptions to the rule that gases freely mix. For example, gases released during production of theatrical fog from dry ice (carbon dioxide) or liquid nitrogen are cold. This means the gas is denser and heavier than air (the molecules are closer together), and they will stay near the floor for a time.

Gases vary greatly in toxicity and flammability. Some are neither flammable nor toxic. This includes all the inert gases, like nitrogen from fog effects and argon from welding. Inert gases like these can only harm you if the amounts in the air are so high that they reduce the amount of oxygen available for you to breathe. Such gases are called "asphyxiants."

Other gases can have toxic effects. They can be irritating, acidic, caustic, poisonous, and so on. Examples of toxic gases encountered in theatrical shops and locations include formaldehyde outgassing from plywood and fabrics, carbon monoxide from fuel-powered lifts running indoors, sulfur dioxide from pyrotechnics, and acetic acid gas from photographic developing baths.

Vapors

Vapors are the gaseous form of liquids. For example, water vapor is created when water evaporates—that is, releases vapor molecules into the air. Once released into the air, vapors behave like gases and expand into space. However, at high concentrations, they will recondense into liquids. That is what happens when it rains.

There is a common misconception that substances do not vaporize until they reach their vaporization point—that is, their boiling point. Although greater amounts of vapor are produced at higher temperatures, most materials begin to vaporize at varying rates as soon as they are in liquid form.

A few solids convert to their vapor form at room temperature. Solids that do this are said to sublime. Mothballs are an example of a chemical solid that sublimes to produce a vapor.

Vapors, like gases, may vary greatly in toxicity and flammability. Common toxic vapors created at work might include organic chemical vapors from solvents, such as paint thinners, lacquer thinners, and fabric-cleaning solvents.

Mists

Mists are tiny, liquid droplets in the air. Any liquid, including water, oil, or solvent, can be misted or aerosolized. The finer the size of the droplet, the more deeply the mist can be inhaled.

Some mists, such as paint-spray mists, also contain solid material. Paint mist can float on air currents for a time. Then the liquid portion of the droplet will vaporize, and the solid part of the paint will settle as a dust.

A mist of a substance is more toxic than the vapor of the same substance at the same concentration. This results from the fact that when inhaled, the droplets deliver the mist in small, concentrated doses to the respiratory system's tissues. Vapors, on the other hand, are more evenly distributed in the respiratory tract.

Fumes

Laymen commonly use this term to mean any kind of emission from chemical processes. In this book, however, only the scientific definition will be used.

Fumes are very tiny particles, usually created in high-heat operations such as welding, soldering, or foundry work. They are formed when hot vapors cool rapidly and condense into fine particles. For example, lead fumes are created during soldering. When solder melts, some lead vaporizes. The vapor immediately reacts with oxygen in the air and condenses into tiny lead oxide fume particles.

Fume particles are so small (0.01 to 0.5 microns in diameter)[1] that they tend to remain airborne for long periods of time. Eventually, however, they will settle to contaminate dust in the workplace, in the ventilation ducts, in your hair or clothing, or wherever air currents carry them. Although fume particles are too small to be seen by the naked eye, they sometimes can be perceived as a bluish haze from soldering or welding operations.

Fuming tends to increase the toxicity of a substance, because the small particle size enables it to be inhaled deeply into the lungs and because it presents more surface area to lung fluids (i.e., is more soluble).

In addition to many metals, some organic chemicals, plastics, and silica will fume. Smoke from burning organic materials usually contains fumes.

Dusts

Dusts are formed when solid materials are broken down into small particles by natural or mechanical forces. Natural wind and weathering produces dusts from rocks. Sanding and sawing are examples of mechanical forces that produce dusts. And many of the substances we purchase, such as talcum powder, powdered pigments, and dyes, are already in a fine dust form.

The finer the dust, the deeper it can be inhaled into the lungs, and the more toxic it will be. Dusts are classified as "respirable" or "nonrespirable," based on their particle size.

- Respirable dusts are under 10 microns in diameter. They are far too small to be seen with the naked eye and can get deep into the lungs' air sacs. If they are inert, they remain there for life. If they are soluble, they will enter the bloodstream in time periods ranging from minutes to years, depending on their solubility in lung fluids. (Fumes are also respirable particulates.)
- Nonrespirable dusts are 10 to 100 microns in diameter and will settle in other parts of the respiratory system, such as the sinuses and the large and small bronchial tubes. They are caught in the mucus lining of the lungs and are transported by small, hair-like projections (cilia) from the cells lining the lung tissue up to the back of the throat, where they are most often swallowed.

---

[1]A micron is a metric system unit of measurement equaling one millionth of a meter. There are about 25,640 microns in an inch. Respirable fume and dust particles are in the range of 0.01 to 10 microns in diameter.

Smoke

Smoke is formed from burning organic matter. Burning wood and hot-wire-cutting plastics are two smoke-producing activities. Smoke is usually a mixture of many gases, vapors, and fumes. For example, cigarette smoke contains over four thousand chemicals, including carbon monoxide gas, benzene vapor, and fume-sized particles of tar.

EXPOSURE: HOW MUCH IS TOO MUCH?

Standards

Exposure to airborne chemicals in the workplace is regulated in the United States and Canada. The United States limits are called OSHA Permissible Exposure Limits (PELs). The Canadian limits are called Occupational Exposure Limits (OELs).

Most of the U.S. OSHA standards were set in the 1970s or earlier, and OSHA has tried repeatedly and unsuccessfully to upgrade them. The last time OSHA tried to update the PELs was in 1992. A coalition of industries took OSHA to court and got the judges to rule that, before changing any PEL, OSHA must provide complete environmental and economic impact statements on each chemical. The documentation for each one of these statements takes years of research and consists of thousands of pages of data. Since there are over four hundred outdated PELs, industry has successfully halted any rational approach to workplace air-quality regulation.

Fortunately, there are other standards that are more reliable than OSHA's and that are regularly reviewed and updated. These are the Threshold Limit Values set by the American Conference of Governmental Industrial Hygienists (ACGIH).

In this book, the TLVs will be used predominantly, with only occasional references to the PELs. The PELs are based on the TLV concept, so that all of the technical information below that applies to TLVs also applies to PELs. Copies of the Threshold Limit Values can be obtained from the ACGIH (see Bibliography).

What Are TLVs?

TLVs are the amounts of chemicals in the air that *almost all healthy adult workers* are predicted to be able to tolerate without adverse effects. There are three types:

- *Threshold Limit Value–Time Weighted Average (TLV-TWA).* These are airborne concentrations of substances averaged over eight hours. They are meant to protect from adverse effects those workers who are exposed to substances at this concentration over the normal eight-hour day and a forty-hour workweek.

- *Threshold Limit Value-Short-Term Exposure Limit (TLV-STEL).* These are fifteen-minute average concentrations that should not be exceeded at any time during a workday.
- *Threshold Limit Value-Ceiling (TLV-C).* These limits are concentrations that should not be exceeded during any part of the workday exposure.

The most commonly used TLVs are the eight-hour time-weighted average limits. Unless the initials "STEL" or "C" appear after "TLV," it is the eight-hour TLV-TWA that is meant.

Do All Chemicals Have TLVs?

One of the greatest problems we face in the workplace is that the majority of the chemicals to which we are exposed have not been tested enough to even determine what levels are safe. Experts estimate there are over seventy thousand chemicals used in the workplace. There are TLVs for only about 700 of these! And new chemicals are being invented every year. Chemicals without TLVs should never be considered safe.

Are TLVs "Safe" Limits?

Exposure at or near the TLV should never be considered proof that any individual worker is safe and cannot sustain damage from that exposure. From the definition of the TLV alone, it is clear that even *a few* "healthy adult workers" will be unable to tolerate concentrations at the TLV.

TLVs also do not apply to people with certain health problems, such as allergies and lung diseases, people taking certain medications or drugs, people who work longer than eight hours per day, children, or fetuses. They need much lower exposures. For these people, the EPA outdoor air-quality standards are probably a better choice. They generally are very much lower than TLVs.

How Do I Use TLVs?

It's easy: the lower the TLV, the less is allowed in the workplace air, and the less you should inhale. On this basis, safer substitutes can be chosen by selecting products with higher TLVs. However, you must also consider the evaporation rates. For example, odorless paint thinner and turpentine have somewhat similar evaporation rates, so it would be wise to use odorless paint thinner, because it has a higher TLV. Acetone, however, is extremely volatile, and while it is less toxic to inhale, you need to consider that you may be inhaling a far larger amount while you work than other solvents.

TABLE 8.1

EXAMPLES OF TLVs IN ORDER OF INCREASING HAZARD

DUSTS & FUMES (solid particles)	TLV-TWA (milligrams/meter³)
nuisance dusts (e.g. plaster or chalk)	10
graphite	2
quartz (e.g., fine dust from sand)	0.1
lead (e.g., lead/chrome pigments, lead solder fume)	0.05
cadmium (e.g., cadmium pigments)	0.005

GAS or VAPOR (molecules in air)	TLV-TWA (parts/million-ppm)
ethyl alcohol (grain alcohol)	1000
acetone	500
odorless paint thinner	300
turpentine	100
n-hexane (e.g., rubber cement thinner)	50
carbon tetrachloride	5
phosgene (chemical warfare gas)	0.1
diisocyanates (from urethane resins/foams)	0.005

When Has the TLV Been Exceeded?

Expensive air sampling and analysis are usually required to prove that Threshold Limit Values are exceeded. OSHA requires this kind of air sampling for any workplace where highly toxic substances, such as lead, cadmium, and arsenic, are used regularly. A good reason to eliminate these highly toxic substances from our pigments, solders, and other materials is that we are technically using them illegally if the employer has not monitored our exposure during work to establish the levels at which we are exposed.

Since personal air monitoring is not often done in our business for all the different substances we use, TLVs are primarily useful as proof that a substance is considered toxic and that measures should be taken to limit exposure to substances with TLVs.

Home Studios: High Exposures

The fact that TLV-TWAs only apply to an eight-hour-per-day exposure is especially relevant when assessing the risk for people who routinely work longer than eight hours or for people who live and work in the same environment, such as artists who work at home. Home-working artists may receive very high exposures, since they may be exposed to contaminants twenty-four hours a day. With no respite during which the body can detoxify, even low concentrations of contaminants become significant.

9 Respirators

This chapter covers proper use of respiratory protection and the OSHA regulations on respiratory protection (29 CFR 1910.134). The information covered in chapter 8, The Air We Breathe, gives you an understanding of the nature of the substances that respirators are designed to capture.

TYPES OF RESPIRATORS

Respirators come in three basic types:

- *Air-purifying respirators.* These use the wearer's breath to draw air though cartridges or filters inserted in a full- or half-faced respirator or through a mask that functions as a filter.
- *Air-powered respirators.* These have a powered unit that mechanically draws air through a filter and delivers it to the interior of a respirator, a hood, or under a faceshield.
- *Air-supplied respirators.* These have systems for bringing fresh air to the wearer by means of pressurized tanks or compressors.

DUST MASKS ARE RESPIRATORS

First, let's correct a common misconception that dust masks are exempt from OSHA rules. They are not. OSHA considers masks as simply one type of respirator and defines them (in 1910.134(b)) this way:

> **Filtering facepiece** (dust mask) means a negative pressure particulate respirator with a filter as an integral part of the facepiece or with the entire facepiece composed of the filtering medium.

The only dust masks not covered by OSHA are the pollen dust masks that can be purchased at most drugstores and some loose-fitting masks used in the medical profession. These clearly state on each mask or on the box that they do not provide protection against toxic substances.

WHEN SHOULD RESPIRATORS BE USED?

Adequate ventilation, not respirators, should be the primary means of controlling airborne toxic substances. OSHA regulations forbid using respirators for primary protection, except when ventilation is being installed, maintained, or repaired, during emergencies, or if engineering controls are shown not to be feasible.

OSHA REGULATIONS

If you regularly wear a respirator on the job, OSHA requires your employer to provide:

- *A written program* explaining how the employer will meet all the requirements below.
- *A written hazard evaluation* to determine hazards you face on the job and the employer's rationale for selecting particular respirators. This often requires an industrial hygienist to come in and do personal air monitoring of workers' exposure during hazardous jobs.
- *A medical evaluation* to determine your ability to wear the selected respirator. The evaluation may be done by either a physician or a "licensed medical professional."
- *Formal fit testing* at least annually by a qualified person using one of the approved methods.
- *Documented training* annually to ensure that you are familiar with the use and limitations of the equipment; procedures for regular cleaning, disinfecting, and maintenance of all respirators; how to don, doff, and do a seal check before each use; and other technical matters. Retraining should be done at least annually.
- *Periodic program evaluation* to ensure that respirator use continues to be effective.

Sadly, these regulations are not being met in most permanent shops and venues. And they certainly are not followed at all on short-term jobs, such as for commercials or film location work.

FITTING RESPIRATORS

You cannot know whether or not your respirator is protecting you unless you have been fit tested by a qualified person. If you are employed, this qualified person must be provided by your employer. If you are self-employed, you will have to arrange your own fit testing. The companies that sell respirators rarely provide fit testing. You can hire an industrial hygienist or consultant, but this is expensive. Some workers mistakenly think they can test themselves by doing a fit check.

Fit Check

Confusion between the terms "fit test" and "fit check" caused OSHA to change the term "fit check" to "user seal check."

A user seal check for a mask is done by putting it on and seeing if it will briefly maintain negative pressure when you inhale or positive pressure when you exhale. You know it doesn't fit if you feel air escaping near your nose, under your chin, or from some other place where the seal is broken.

To perform a user seal check for a cartridge respirator, you need to close the exhalation valve with your hand and exhale into the facepiece. Next, you want to block air coming into the cartridges on the inhale with your hands. The facepiece will not let air leak in or out on either procedure.

Employers are required to provide trainers who will give workers hands-on instruction on seal checks. Workers are expected to do this check every time they put on a respirator or mask. *But this seal check is not a fit test.*

The Real Fit Test

Employers are required under the OSHA regulation to provide fit testing of all respirators (including masks). Your employer may hire a consultant or have an employee specially trained to do the job. The regulations describe the various approved methods for doing the two basic types of fit testing.

- *Qualitative fit testing* depends on the wearer's ability to sense an odor, taste, or irritation from one of four approved chemicals delivered in a controlled way to an enclosure around the user's head. Chemicals approved for this use include banana oil (isoamyl alcohol), saccharine mist, an irritant smoke (e.g., stannic chloride), and a bitrex (e.g., denatonium benzoate). These tests are not allowed for full-face negative pressure and supplied air pressure demand respirators.
- *Quantitative fit testing* is done by measuring and numerically comparing the contaminants inside and outside the mask or respirator. The equipment needed to do this is expensive, but the tests are appli-

cable to all types of respirators, they are more accurate, and they create a document that makes it easy for employers to keep on file the required written records.

Beards

Many men mistakenly think that they can successfully wear their respirator over a beard. They can't. The new OSHA rules contain very explicit wording about facial hair. Under 29 CFR 1910.134(g), *Use of respirators,* it says:

> (1) *Facepiece seal protection.* (i) The employer shall not permit respirators with tight-fitting facepieces to be worn by employees who have:
>
> (A) Facial hair that comes between the sealing surface of the facepiece and the face or that interferes with valve function,
>
> or
>
> (B) Any condition that interferes with the face-to-facepiece seal or valve function.

In addition, in the mandatory fit testing procedures in Appendix A, Part I, it says:

> 9. The [fit] test shall not be conducted if there is any hair growth between the skin and the facepiece sealing surface, such as stubble beard growth, beard, mustache, or sideburns which cross the respirator sealing surface. . . .

Employers who let their bearded workers wear respirators risk OSHA citations. In fact, employers can require their workers to shave.

Saving the Beard

A hooded supplied-air system will work for most workers who are determined to keep their beards. These systems are expensive and require the purchase of a special compressor that can produce air that is safe to breathe. Ordinary shop compressors must not be used for this purpose. The complete supplied-air system usually costs well over $1,000.

CHOOSING THE RIGHT CARTRIDGE OR FILTER

Standards

The acronym "NIOSH" is always somewhere on masks and respirators. This stands for the National Institute for Occupational Safety and Health. NIOSH specifies the tests that the products of all manufacturers of respirators must pass. There are NIOSH standards for both the chemical and the particulate types of filters.

In 1998, OSHA rewrote the respiratory protection standard. Of all the changes, the most important for artists was the adoption of new standards for the filters that are in masks or in respirator cartridges.

Chemical Cartridges

Chemical cartridge standards have not changed. These filters capture substances such as ammonia, formaldehyde, acids, and organic solvents. It is crucial to determine precisely what is in the air in order to pick the right cartridge (see table 9.1 below). Use of the wrong one will render the respirator useless for the purpose.

TABLE 9.1 CHEMICAL CARTRIDGE TYPES

CARTRIDGE	COLOR CODE	CONTAMINANT
acid gas	white	gases rising from acids, bleaches, and some photochemicals
formaldehyde	khaki	plywood, urea formaldehyde glues
ammonia	green	cleaners, diazo copier chemicals
organic vapor	black	evaporating solvents
organic vapor/acid gas	yellow	both organic vapors and acid gases
mercury/chlorine	orange	mercury vapors and chlorine gas
organic vapor	black	paint, lacquer and enamel spray
P100 particulate	purple	airbrush, aerosol paints

There are also many substances for which there are no approved chemical cartridges. For a partial list of these substances, see the section below, "When Air Supplied Systems are Needed."

Particulate Filters

Particulate filters are those that capture airborne solid particles and liquid mists. These are the ones artists can use to protect themselves from clay and glaze dusts and glaze spray mists. Picking the right mask for artwork used to be easy. The filters came in only three types: dust, mist, and fume. Now, there are nine different types. And we should know them.

Performance Classifications. The nine types of particulate filters now come in three different series, designated as N, R, and P. All of these filters are tested against fume-sized particles (0.3 microns). The filters in each series have three minimum efficiency levels: 95 percent, 99 percent, and 99.97 percent. That is:

- N95, R95, and P95 filters are certified as having a minimum efficiency of 95 percent.
- N99, R99, and P99 filters are certified as having a minimum efficiency of 99 percent.
- N100, R100, and P100 filters are certified as having a minimum efficiency of 99.97 percent.

N series filters can be used only in atmospheres containing non-oil-based particulates. For example, you should not use the N filters if people nearby are machining metal with cutting oils or spraying WD40®.

The N filters also have a time-use restriction. This means you should only use them for eight hours. The eight hours can be either continuous or intermittent. Intermittent use means that you are using the filter short periods of time, sealing it in plastic Ziploc® bags between wearings, and will discard the mask when the amount of time it was used adds up to eight hours.

R series filters can be used whether or not there is oil present. R filters also have a time-use restriction of eight hours of continuous or intermittent use.

P series filters may be used in either a non-oil- or oil-containing atmosphere. They do not have any time-use restrictions, which means they can be used until they are soiled, damaged, cause an increase in breathing stress, or show some other sign that they are worn out. The P100 is the top of the line and the only filter to be assigned the familiar magenta color reserved for the old HEPA or fume filter.

Selection Criteria

Since it is rare that oil mists are present in the studio, we may be able to use any of these filters. To choose the right one, you need to consider two primary factors: the toxicity and the particle size of the contaminant.

Toxicity. The choice is simple if you are working with highly toxic substances, such as powdered pigments containing lead, cadmium, or chrome. In these cases, you always use a filter with a 100 percent rating. For example, OSHA requires a 100 percent HEPA filter be used whenever lead is airborne, regardless of particle size.

Particle Size. Silica flour, metal fumes, pigment powders, and other substances known to contain significant amounts of particles in the range of 0.3 microns in diameter require a respirator with a filter efficiency of 100. Unfortunately, some pigment and dye products are smaller in size than 0.3 microns, and perfect capture cannot be assumed even with the very best of filters.

Course grinds or granular materials of moderate or low toxicity may be addressed with one of the less efficient filters. It would be easier to choose a respirator if suppliers of powdered materials would provide the particle size distribution data on their products. This data is readily available from most of the primary manufacturers, but secondary product manufacturers usually will not pass it along to you.

Until you know the particle size of the materials you use, it is wise to use the 100 filters. You might want some 95s around for wood dust, plaster dust, or other large particle and low toxicity dusts.

Choosing whether you want an N100, R100, or P100 mask will depend on how you work. If, at the end of the day, your mask usually looks dirty and bent out of shape, you should buy the least expensive N series mask and replace it frequently. If your masks usually look as good as new at the end of the day, you will save money buying the more expensive P100s and taking good care of them.

Protection Factors. The choice of a respirator generally involves selecting among disposable masks and half- and full-faced respirators. Disposable masks are designed for lighter exposures (up to five times the TLV) than respirators (up to ten times the TLV). Respirators with large-capacity canisters are designed for heavily contaminated air. (To handle more heavily contaminated atmospheres, use air-supplied respirators.) Use full-face respirators when the contaminant is also an eye hazard or irritant.

WHEN AIR-SUPPLIED SYSTEMS ARE NEEDED

Air-supplied, rather than air-purifying, respirators should never be used in oxygen-deficient atmospheres, such as when gas is released in a confined space or in firefighting. OSHA defines oxygen-deficient atmospheres as 19.5 percent oxygen or less. Since air ideally contains about 21 percent oxygen, it is easy to see that it only takes a few percentage points of another gas in the air before this is met. While air containing only 19.5 percent oxygen is not immediately life-threatening, it is not an atmosphere conducive to hard work and heavy respiration.

Air-supplied respirators should also be used against chemicals that are of an extremely hazardous nature, that lack sufficient warning properties (smell or taste), that are highly irritating, or that are not effectively absorbed by filter or cartridge material. Included among these are hot or burning wax vapors (acrolein and other hazardous decomposition products), carbon monoxide, methyl (wood) alcohol, isocyanates (from foaming or casting polyurethane), nitric acid, ozone, methyl ethyl ketone peroxide (used to harden polyester resins),and phosgene gas (created when chlorinated solvents come into contact with heat or flame).

RESPIRATOR CARE

At the end of a work period, clean a respirator and store it out of direct sunlight in a sealable plastic bag. Respirators should never be hung on hooks in the open or left on counters in the shop. If a respirator is shared, it should be cleaned and disinfected between users and the cartridges changed. Inspect respirators carefully and periodically for wear and damage each time you put it on.

WHERE TO BUY RESPIRATORS

The yellow pages of any good-sized city's telephone book will list distributors under the heading "Safety Equipment." Or consult *Lab Safety and Supply/Other Suppliers* (see Bibliography).

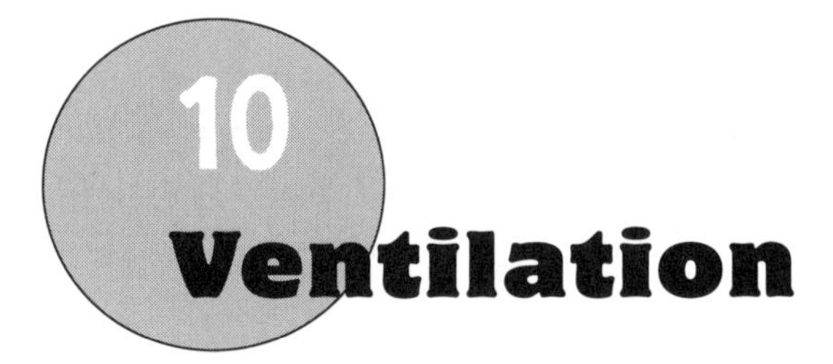

10 Ventilation

Whenever toxic materials must be used, methods for reducing exposure to these materials must be employed. Two methods are respiratory protection, which was discussed in the previous chapter, and ventilation. OSHA requires that ventilation, rather than respiratory protection, be used whenever feasible. And there are rarely any circumstances under which ventilation is not feasible.

TYPES OF VENTILATION

Without ventilation, people will not remain healthy indoors even if they are only doing office work! To keep people healthy in buildings, there are two basic types of ventilation:

- *Comfort ventilation* to provide sufficient air for health and comfort in nonindustrial work environments, including theater lobbies, ticket offices, and the house
- *Industrial ventilation* to protect people working with toxic substances that become airborne when working in shops, cleaning, renovating, and the like

Comfort Ventilation

Comfort ventilation provides sufficient air movement and fresh air to avoid buildup of small amounts of air pollution from sources such as chemicals released by furniture, carpets, or building materials, office machinery, and the carbon dioxide expelled when people breathe. This is usually accomplished by either natural or recirculating ventilation systems.

Natural ventilation takes advantage of rising warm air and prevailing winds to cause the air to circulate and exchange with outside air in sufficient amounts to provide comfort. Often, chimney-like flues are constructed behind walls to draw out warm air, and windows either leak air or can be opened. Such systems are found only in very old theaters, where high ceilings and open spaces enhance the system.

Recirculating ventilation systems use fans or blowers to circulate air through ducts from room to room throughout the building. On each recirculating cycle, some fresh air from outside is added and some recirculated air is exhausted. The amount of fresh air added commonly varies from 5 to 30 percent, depending on the vagaries of the particular ventilation system or its operator. Building managers often are encouraged to add as little fresh air as possible to reduce heating and cooling costs.

Air-conditioning. Whether it is a centrally located unit for an entire building or a small room window unit, air conditioners do not provide sufficient fresh air for health. Air conditioners draw in air, cool it, and return it. They may remove some humidity, but they do not remove toxic gases and vapors. Dusts of large diameter, such as some pollens, may be removed by air-conditioning filters, but respirable dust particles, gases, and vapors will pass through easily. Some window air conditioners have an exhaust setting, but the amount of air exhausted usually is not sufficient to provide good comfort ventilation, and certainly not enough to protect people from toxic pollutants.

Sick building syndrome occurs when insufficient amounts of fresh air are added to comfort ventilation systems. People in these buildings may complain of eye irritation, headaches, nausea, and other symptoms. Taken together, these are sometimes called "sick building syndrome." The symptoms are apparently caused by the accumulation of body heat, humidity, dust, molds and other microorganisms, formaldehyde, and other common air pollutants.

Standards. The American Society of Heating, Refrigerating, and Air-conditioning Engineers' (ASHRAE) most recent standard for indoor air quality, 62-1999, says:

> The purpose of the standard is to specify minimum ventilation rates and indoor air quality that will be acceptable to human occupants and are intended to minimize the potential for adverse health effects.

In order to avoid adverse health effects, ASHRAE 62-1999's minimum rates are, in most instances, fifteen or twenty cubic feet per minute per person outside fresh air (not recirculated) delivered to the level at which people are breathing (not supplying and exhausting air at the ceiling, as most recirculating systems do).

ASHRAE standards apply to theater houses, box offices, lobbies, and other public and office areas. *ASHRAE standards never apply to shops which use toxic substances.* This is explained in ASHRAE 62-1999's scope:

> This standard applies to all indoor or enclosed spaces that people may occupy, except where other applicable standards and requirements dictate larger amounts of ventilation than this standard.

These larger amounts of air to which they refer are those required for removal or dilution of toxic substances by the standards of the American Conference of Governmental Industrial Hygienists (ACGIH). The larger amounts of air are required to keep workers' exposures below the threshold limit values (TLVs). To accomplish this, it is necessary to provide ventilation that is consistent with ACGIH standards for *industrial* ventilation.

INDUSTRIAL VENTILATION

ACGIH sets internationally accepted standards for workplace air quality and for industrial ventilation in their publication, *Industrial Ventilation: A Manual of Recommended Practice* (see Bibliography). ACGIH sets standards for both basic types of industrial ventilation: *dilution* ventilation and *local exhaust* ventilation.

Dilution ventilation does exactly what its name implies. It dilutes or mixes contaminated workplace air with large volumes of clean air to reduce the amounts of contaminants to acceptable levels. Then, the diluted mixture is exhausted (drawn by fans or other devices) from the workplace.

Dilution systems usually consist of fresh-air inlets (often having fans and systems for heating or cooling the air) and outlets (exhaust fans). The fresh replacement (makeup) air inlets should be placed as far from the exhaust fan as possible and in a location designed to take advantage of the flow of air from inlet to exhaust (figure 10.1).

Although often cheap and easy to install, dilution ventilation has limited uses. For example, only vapors or gases of low toxicity or very small amounts of moderately toxic vapors or gases are removed sufficiently by dilution ventilation. *These systems should not be used for control of dusts, mists, or highly toxic air contaminants in any form.*

Local exhaust ventilation is the best means by which large amounts of airborne substances, or substances of moderate to high toxicity, are removed from the workplace. Table 10.1 lists processes that require local exhaust ventilation.

Local exhaust systems consist of 1) a hood enclosing or are positioned very close to the source of contamination to draw in the air; 2) ductwork to carry away the contaminated air; 3) if needed, an air cleaner to filter or purify the air before it is released outside; and 4) a fan to pull air through the system. Because local exhaust ventilation captures the contaminants at their source rather than after they have escaped into the room air, exhaust ventilation systems remove smaller amounts of air than dilution systems. This means that these systems cost less to run, because less replacement air must be heated, cooled, or air-conditioned.

Types of hoods. A hood is the structure through which the contaminated air first enters the system. Hoods can vary from small dust-collecting types built around grind wheels to two-story-high spray booths for theatrical scenery. The drawings and specifications for every type of hood needed in the university setting can be found in the ACGIH *Manual of Recommended Practice.* Some specific hood types for various processes are shown in figure 10.2.

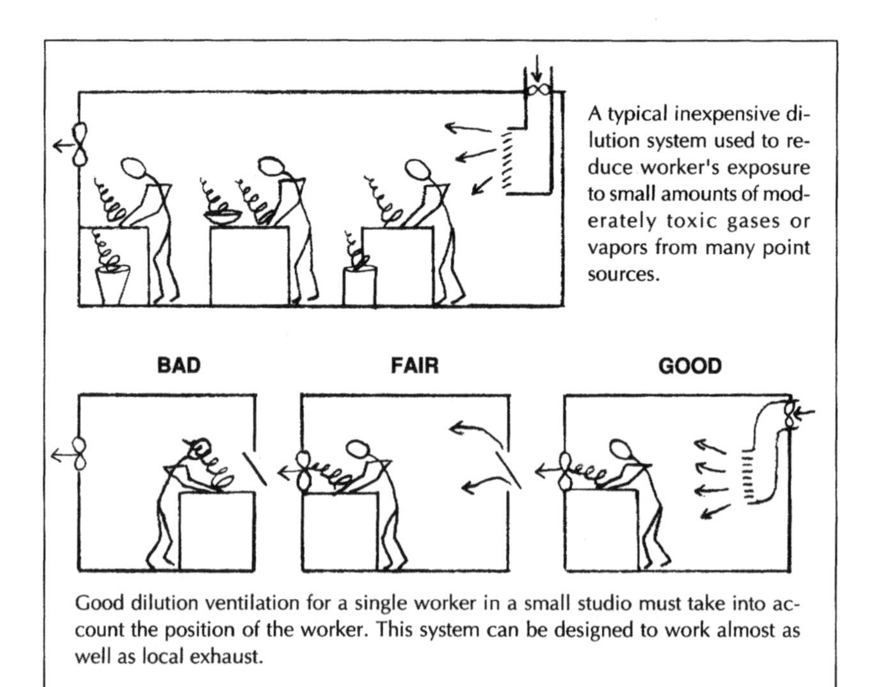

FIGURE 10.1 DILUTION VENTILATION

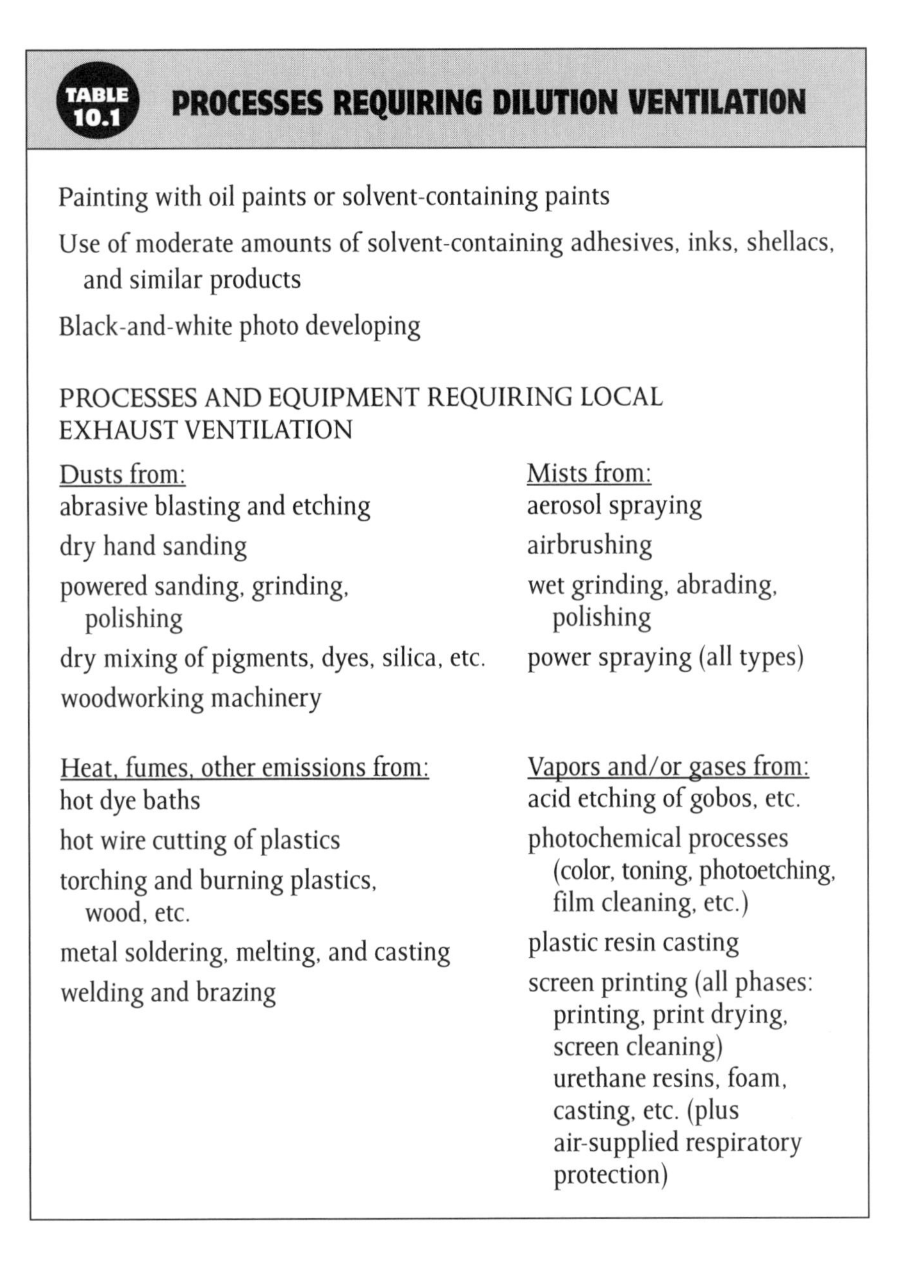

TABLE 10.1 PROCESSES REQUIRING DILUTION VENTILATION

Painting with oil paints or solvent-containing paints

Use of moderate amounts of solvent-containing adhesives, inks, shellacs, and similar products

Black-and-white photo developing

PROCESSES AND EQUIPMENT REQUIRING LOCAL EXHAUST VENTILATION

Dusts from:	Mists from:
abrasive blasting and etching	aerosol spraying
dry hand sanding	airbrushing
powered sanding, grinding, polishing	wet grinding, abrading, polishing
dry mixing of pigments, dyes, silica, etc.	power spraying (all types)
woodworking machinery	

Heat, fumes, other emissions from:	Vapors and/or gases from:
hot dye baths	acid etching of gobos, etc.
hot wire cutting of plastics	photochemical processes (color, toning, photoetching, film cleaning, etc.)
torching and burning plastics, wood, etc.	plastic resin casting
metal soldering, melting, and casting	screen printing (all phases: printing, print drying, screen cleaning) urethane resins, foam, casting, etc. (plus air-supplied respiratory protection)
welding and brazing	

Types of ductwork. One striking difference between comfort and industrial ventilation is that comfort systems usually employ square ducts. *Industrial ventilation almost always employs round ducts.* One of the major reasons is that the high duct velocities necessary for keeping particulates in suspension (~ 4,000 feet per minute) cannot be accommodated by a square duct without great loss of efficiency and incredible noise.

TYPES OF LOCAL SYSTEMS

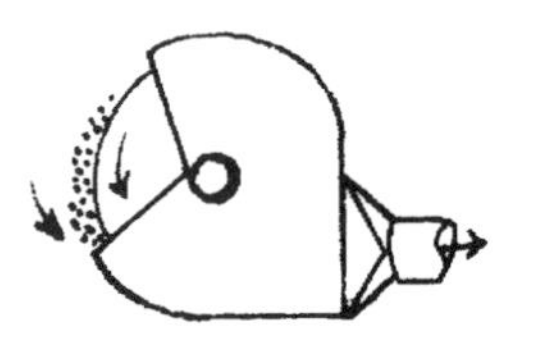

[FIGURE 10.2]

Dust-collecting systems. Most grind wheels, table saws, and other dust-producing machines sold today have dust-collecting hoods built into them. Some machines need only to be connected to portable dust collectors that can be purchased off the shelf. In other cases, stationary ductwork can be used to connect machines to dust collectors such as cyclones (which settle out particles) and baghouses (which capture particles on fabric filters).

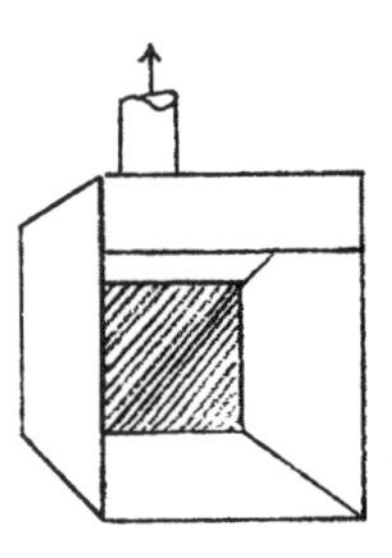

[FIGURE 10.3]

Spray booths. Spray booths from small table models to walk-in–sized or larger can be purchased or designed. Spray booths can be used for processes such as spraying of paints, lacquers, adhesives, and other materials, plastic resin casting, paint stripping, and solvent cleaning. If the materials used in the booth contain flammable solvents, the spray booth, its ducts and fans, and the area surrounding the booth must be made safe from explosion and fire hazards, as required by the General Industry Standards and other local codes.

[FIGURE 10.4]

Movable exhaust systems. Also called "elephant trunks," these flexible duct and hood arrangements are designed to remove fumes, gases, and vapors from processes such as welding, soldering, or any small, tabletop processes that use solvents or solvent-containing products. Movable exhausts can also be equipped with pulley systems or mechanical arms designed to move hoods to almost any position.

Canopy hood systems. These hoods take advantage of the fact that hot gases rise. They are used over processes such as hot dye baths, wax and glue pots, stove ranges, and the like. Unfortunately, they are installed above some theater shop worktables where they are not only ineffective, because the hood is too far from the table, but where they are even dangerous, because they draw contaminated air past the worker's face.

[FIGURE 10.5]

Slot hood systems. These systems draw gases and vapors across a work surface, away from the worker. Slot hood systems are good for any kind of bench work.

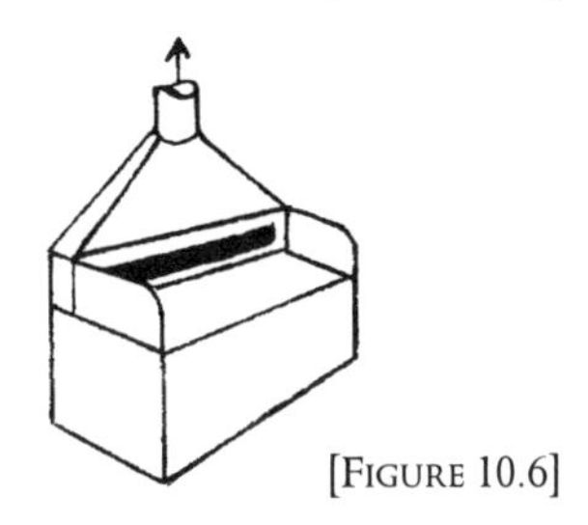
[FIGURE 10.6]

PROVIDING CLEAN AIR

Standards

The purpose of ventilation is to meet air-quality standards. Standards for industrial air quality are set by ACGIH. TLVs are determined by ACGIH for about 700 toxic air contaminants. The TLVs are designed to protect *almost all healthy adult employees*. Workers who are physically unable to tolerate chemical exposure at the TLVs often must be accommodated by employers.

The TLVs apply only to work in the shops. In the house, air quality must be far better. This is where audience members, people with chronic illnesses or sensitivities, and children may be present. The general public, unlike employees, have not tacitly agreed, by virtue of being employed, to accept the risk that exposure at the TLV carries. They have a right to much more protection and better quality air.

Air Purifiers

Any device that draws in air, filters it, and returns the air to the room should be regarded with suspicion. These devices are very limited in application. The following are some familiar examples.

- *Negative ion and/or ozone generators* are useless, and many are even dangerous.

- *Electrostatic precipitators (ESPs)* cannot capture gases or vapors. They capture only certain types of particulates. For example, ESPs used in welding operations will capture metal fumes, but cannot collect toxic welding gases, such as ozone and nitrogen oxides.
- *Charcoal filtered air purifiers* will collect certain gases and vapors. Usually, the filters are costly and must be changed far more frequently than is practical.
- *HEPA* (high efficiency particulate air) filters will capture small dust and fume particles. These filters must not be used in ceiling or wall units if their fans create turbulence, so that they stir up as much dust as they collect. They are often useful in wood shops to collect the fine dusts that escape from other dust control systems, such as baghouses.

DESIGNING OR RENOVATING VENTILATION SYSTEMS

Steps

1. First, choose an industrial hygienist to interview employers and workers in order to specify systems that will provide proper ventilation within the budget.
2. Select a professional engineer who is *experienced in industrial ventilation* to modify and adapt the chosen ventilation system designs found in the ACGIH publication, *Industrial Ventilation: A Manual of Recommended Practice.*
3. For other areas where no toxic substances are airborne, a heating, ventilating, and air-conditioning (ASHRAE) engineer will be needed to help integrate the new industrial systems so that they will not unbalance comfort ventilation systems.
4. Review blueprints of all plans for new construction or renovation of ventilation systems. Ask designers to explain the systems in detail before construction begins.
5. Pick a contractor who understands the differences in the two types of systems. These contractors will follow the plans and will not be tempted to try to save money by making unauthorized modifications.
6. When the system is installed, have it checked by an industrial hygienist or other expert *who has nothing to gain or lose financially from approving the system.*

Liability

Avoid using architects and engineers who mistakenly think they only need to meet local building codes and laws. Their work must also meet ASHRAE and ACGIH professional standards of practice. I have personally participated in a legal action that was settled by payment of large sums by architects and engineers who designed an art building that did not meet ACGIH industrial ventilation standards.

Defining System Limits

A written policy should be developed for each new installation that clearly defines what the system can and cannot do. Any activity other than those listed in the policy would require written approval of building engineers or others familiar with both the proposed process and the ventilation system's efficacy.

Ventilation systems are only able to do the job for which they were designed. Processes that cannot be safely done in an existing system must be prohibited in writing. Improper use of good systems most often occurs when a newly hired scenic artist wants to use more hazardous materials. The shop must have a firm policy that prohibits new procedures unless the ventilation is proven to be adequate. Safety must come before any artistic consideration.

Is the Ventilation Working?

Even without the advice of an expert, some commonsense observations can be made about local exhaust systems in the shop:

- Can you see the system pulling dusts and mists into it? If not, you might use incense smoke or soap bubbles to check the system visually. When released in the area where the hood should be collecting, the fog, smoke, or bubbles should be drawn quickly and completely into the system.
- Can you smell any gases or vapors? Sometimes placing inexpensive perfume near a hood can demonstrate a system's ability to collect vapors, or it can show that exhausted air is returning to the workplace (or some other place where it should not be).
- Do people working with the system complain of eye, nose, or throat irritation or have other symptoms?
- Is the fan so noisy and irritating that people would rather endure the pollution than turn it on? Fans should not be loud, and experts should be expected to work on the system until it is satisfactory.

- Is the system being maintained? There should be written maintenance schedules for ventilation systems. If there are none, expect that the system is not—or soon will not be—functioning properly. If there are schedules, check to see how often filters have been changed, ducts cleaned, fan belts replaced, and the like.

SPECIAL VENTILATION FOR THE HOUSE

Theaters themselves also have special ventilation requirements that involve both comfort and industrial systems. For example, the house needs general air-conditioning, and there must also be special smoke-exhaust fans in the fly to protect the audience areas during a fire.

Theater air-conditioning should be designed so that air flows from the house to the back of the stage, to keep special effects, such as smoke and fog, out of the audience. A method of flushing the stage area to remove special effects emissions between acts is also needed. There also are special local-exhaust air-curtain systems that have been designed for the orchestra pit in some theaters to keep special effects emissions away from the musicians.

The ventilation systems in theaters must be engineered specially to ensure that they are very quiet in order not to distract audience attention. And comfort ventilation for the house is not only a health issue, it is a business issue. Audiences that are too hot, too cold, or must endure stale odors will not be easy to entertain. People also become inattentive and fall asleep easily if insufficient fresh air is introduced and carbon dioxide levels rise.

11 Solvents

Solvents are used in most paints (including many so-called water-based paints), varnishes, inks, and their thinners, in aerosol spray products, leather and textile dyes, permanent marking pens, glues and adhesives, some photographic chemicals, and much more.

WHAT ARE SOLVENTS?

The term "solvents" refers to liquid, organic chemicals used to dissolve solid materials. Solvents can be made from natural sources, such as turpentine and the citrus solvents, but most are derived from petroleum or other synthetic sources. Solvents are used widely, because they dissolve materials like resins and plastics, and because they evaporate quickly and cleanly.

SOLVENT TOXICITY

There are no safe solvents. All solvents, natural or synthetic, are toxic. Contact with liquid solvents or inhalation of the vapors they emit into the air are both hazardous.

In general, solvents can irritate and damage the skin, eyes, and respiratory tract, cause a narcotic effect on the nervous system, and damage internal organs such as the liver and kidneys. These kinds of damage can be acute (from single heavy exposures) or chronic (from repeated low-dose exposures over months or years). In addition, some solvents are especially hazardous to specific organs or can cause specific diseases, such as cancer.

Skin Contact

All solvents can dissolve the skin's protective barrier of oils, drying and chapping the skin and causing a kind of dermatitis. In addition, some solvents can cause severe burns and irritation. Natural solvents such as turpentine and limonene are known to cause skin allergies. Other solvents may cause no symptoms, but may penetrate the skin, enter the bloodstream, travel through the body, and damage other organs.

The Eyes and Respiratory Tract

All solvent vapors can irritate and damage the sensitive membranes of the eyes, nose, and throat. Inhaled deeply, solvent vapors can also damage lungs. The airborne concentration at which irritation occurs varies from solvent to solvent. Often, workers are unaware of solvents' effects at low concentrations. Their only symptoms may be increased frequency of colds and respiratory infections. Years of such exposure could lead to lung diseases, such as chronic bronchitis.

At higher concentrations, symptoms are more severe and may include nosebleeds, running eyes, and sore throat. Inhaling very high concentrations or aspirating liquid solvents may lead to severe disorders, including chemical pneumonia and death. Liquid solvents splashed in the eyes can cause eye damage.

The Nervous System

All solvents can affect the brain or central nervous system (CNS), causing narcosis. Immediate symptoms of this effect on the CNS may include dizziness, irritability, headaches, fatigue, and nausea. At progressively higher doses, the symptoms may proceed from drunkenness to unconsciousness and death. Years of chronic exposure to solvents can cause permanent CNS damage, resulting in memory loss, apathy, depression, insomnia, and other psychological problems that are hard to distinguish from problems caused by everyday living.

Solvents also may damage the peripheral nervous system (PNS), which is the system of nerves leading from the spinal cord to the arms and legs. The symptoms caused by this PNS damage are numbness and tingling in the extremities, weakness, and paralysis. Some solvents, such as n-hexane (found in some rubber cements and many spray products), can cause a combination of CNS and PNS effects, resulting in a disease with symptoms similar to multiple sclerosis.

Damage to Internal Organs

There is considerable variation in the kinds and degrees of damage different solvents can do to internal organs. Many solvents can damage the liver and kidneys, as these organs attempt to detoxify and eliminate the solvents from the body. One solvent, carbon tetrachloride, has such a devastating effect on the liver, especially in combination with alcohol ingestion, that deaths have resulted from its use. Many solvents can also alter heart rhythm, even causing heart attacks or sudden cardiac arrest at high doses. This may be the mechanism that has killed many glue sniffers.

Some solvents are also known to cause cancer in humans or animals. Benzene can cause leukemia. Carbon tetrachloride can cause liver cancer. Many experts suspect that most of the chlorinated solvents (those with "chloro" or "chloride" in their names) are carcinogens.

Reproductive Hazards and Birth Defects

The reproductive effects of solvents are not well-researched. Those studies that do exist show there is reason for concern. For example, Scandinavian studies show higher rates of miscarriages, birth defects, and other reproductive problems among workers exposed to even relatively low levels of solvents. And now, a Canadian study links solvents with major birth defects. For a more complete discussion of these studies and reproductive hazards, see chapter 4.

EXPLOSION AND FIRE HAZARDS

Two properties that affect a solvent's capacity to cause fires and explosions are *evaporation rate* and *flash point*. In general, the higher a solvent's evaporation rate (see definition in table 11.1), the faster it evaporates and the more readily it can create explosive or flammable air/vapor mixtures.

Flash points are the lowest temperature at which vapors are created above a solvent's surface in sufficient amounts to ignite in the presence of a spark or flame. The lower the solvent's flash point, the more flammable it is. Materials whose flash points are at room temperature or lower are particularly dangerous.

The chlorinated hydrocarbons (see Chemical Classes of Solvents, below) are usually not flammable and have no flash points. However, some can react explosively on contact with certain metals, and heating or burning them creates highly toxic decomposition products, including phosgene gas. Hazardous amounts of these toxic gases can be created even by working with chlorinated solvents in a room where a pilot light is burning. *Clearly, all solvents should be isolated from sources of heat, sparks, flame, and static electricity.*

CHEMICAL CLASSES OF SOLVENTS

All solvents fall into various classes of chemicals. A class is a group of chemicals with similar molecular structures and chemical properties. Important classes of solvents are aliphatic, aromatic, and chlorinated hydrocarbons, alcohols, esters, and ketones. Table 11.1 shows various solvents and their properties by class.

RULES FOR CHOOSING SAFER SOLVENTS

1. *Compare Threshold Limit Values.* Choose solvents with high TLVs whenever possible. (See complete definition in the U.S. data sheet on TLVs.)

2. *Compare evaporation rates.* Choose solvents with low evaporation rates whenever possible. In fact, some very toxic solvents that evaporate very slowly may not be as hazardous to use as less toxic ones that evaporate very quickly.
3. *Compare flash points.* Choose solvents with high flash points whenever possible. Chlorinated solvents with no flash points should not, however, be considered safe. (See Explosion and Fire Hazards above.)
4. *Compare toxic effects.* Although all solvents are toxic, some may be especially dangerous to you. For example, if you have heart problems, it makes sense to avoid solvents known for their toxic effects on the heart.
5. *Compare within classes.* Often, solvents in the same chemical class can be substituted for each other. Due to their similar chemical structures, many will dissolve the same materials or work the same way.

RULES FOR SOLVENT USE

1. *Try to find replacements for solvent-containing products.* New and improved water-based products are being developed. Keep abreast of developments in new materials.
2. *Use the least toxic solvent possible.* Use table 11.1 to select the safest solvent in each class. Consult Material Safety Data Sheets on the products you use, and choose those containing the least toxic solvents.
3. *Insist on compliance with OSHA hazard communication laws* at your workplace (see chapter 5, Labels).
4. *Avoid breathing vapors.* Use solvents in areas where local exhaust ventilation is available. Dilution ventilation should only be used when very small amounts of solvents or solvent-containing products are used (see chapter 10, Ventilation).
5. *Use self-closing waste cans* for solvent-soaked rags, keep containers closed when not in use, and design work practices to reduce solvent evaporation.
6. *Keep a respirator* with organic cartridges or an emergency air-supplying respirator at hand in case of spills or ventilation failure.
7. *Avoid skin contact.* Wear gloves for heavy solvent exposure, and use barrier creams for incidental light exposures. Wash off splashes

immediately with water and mild soap. Never clean hands with solvents or solvent-containing hand cleaners. If solvents in amounts larger than a pint are used at one time, or if large spills are possible, have an emergency shower installed.

8. *Protect eyes from solvents.* Wear chemical splash goggles that meet ANSI standard Z87.1 whenever there is a chance a splash may occur. An eyewash fountain or other approved source of clean water that provides at least fifteen minutes' flow should be available.

9. *Protect against fire, explosion, and decomposition hazards.* Follow all OSHA and local fire codes. Store amounts larger than a gallon in approved flammable-storage cabinets (this recommendation exceeds requirements). Do not use heat and/or ultraviolet light sources near chlorinated hydrocarbons. Ground containers from which solvents are dispensed. Local exhaust ventilation fans for solvent vapors must be explosion-proof.

10. *Be prepared for spills.* Check all applicable local and federal regulations regarding release of solvent liquids and vapors. If spills of large amounts are likely, use chemical-solvent absorbers sold by most major chemical supply houses. Special traps to keep solvent spills out of sewers may be required by law. Release of large amounts of liquid or vapor of certain solvents must be reported to environmental protection authorities.

11. *Use and dispose of solvents in accordance with local or federal regulations.* These vary around the country, depending on the type of sewage treatment systems, air quality problems, and other factors that determine how solvents may be used and discarded. You may need to call a local department of environmental protection, publicly owned water treatment facility, or other governmental agency to find out the rules in your area.

TABLE 11.1 COMMON SOLVENTS AND THEIR HAZARDS

COLUMN 1 Solvent class designates the chemical group into which solvents fall. Under each class heading are listed individual solvents and their common synonyms. Try to use the safest solvent in each class.

COLUMN 2 Threshold Limit Value-Time Weighted Averages are the ACGIH (American Conference of Governmental Industrial Hygienists) eight-hour, time-weighted Threshold Limit Values (TLV-TWA) for 1999 in parts per million (ppm). When no TLV-TWA exists or a more protective standard is available, a TLV-C, an OSHA permissible exposure limit (PEL), or a Workplace Environmental Exposure Limit (WEEL) from the American Industrial Hygiene Association may be substituted, with notation.

COLUMN 3 Odor Threshold (OT) in parts per million (ppm). Often, OTs are given as a range of amounts that normal people can detect. Solvents whose odor cannot be detected until the concentration is above the TLV are particularly hazardous.

COLUMN 4 Flash Point (FP) in degrees Fahrenheit (°F). The FP is the lowest temperature at which a flammable solvent gives off sufficient vapor to form an ignitable mixture with air near its surface. The lower the FP, the more flammable the solvent. Some petroleum solvents exhibit a range of FPs.

COLUMN 5 Evaporation Rate (ER) is the rate at which a material will vaporize (volatilize, evaporate) from the liquid or solid state when compared to another material. The two common liquids used for comparison are butyl acetate and ethyl ether.

WHEN BUTYL ACETATE = 1.0	WHEN ETHYL ETHER = 1.0
> 3.0 = FAST	< 3.0 = FAST
0.8–3.0 = MEDIUM	3.0–9.0 = MEDIUM
< 0.8 = SLOW	> 9.0 = SLOW

9

Material adapted from *The Artist's Complete Health and Safety Guide,* 2nd Ed, 1995, Rossol. Evaporation rates are listed as fast, medium, or slow.

COLUMN 6 Comments about the particular toxic effects of the solvent. Symptoms listed here are in addition to the general solvent hazards common to all solvents, such as skin damage, narcosis, etc.

SOLVENT CLASS	TLV-TWA	OT	FP	ER	COMMENTS
name, synonyms	ppm	ppm	°F		major hazards in addition to general hazards noted in the data sheet
ALCOHOLS					**One of the safer classes.**
ethyl alcohol, ethanol, grain alcohol, denatured alcohol	1,000	49–716	55	MED	Least toxic in class. Denatured alcohol contains small amounts of various toxic additives.
isopropyl alcohol, propanol, rubbing alcohol	400 200 (NIC)	43	53	MED	One of the least toxic. Long-term hazards not fully studied.
methyl alcohol, methanol, wood alcohol	200	4–6000	52	FAST	High doses or chronic exposure causes blindness. Skin absorbs.
n-propyl alcohol, n-propanol	200	0.03–41	59	MED	Causes mutation in cells. Not evaluated for cancer. Skin absorbs.
isoamyl alcohol, 3-methyl-1-butanol, fusel oil	100	0.03–.07	109	SLOW	Irritation begins at the TLV.
n-butyl alcohol, n-butanol	C 50 (Ceiling)	0.1–11	95	SLOW	Respiratory irritation at well below the TLV. Lacrimator. Skin absorbs.
ALIPHATIC HYDROCARBONS					**Many of these are mixtures of chemicals derived from petroleum.**
kerosene	none	unk	100–150	VERY SLOW	Low toxicity. Aspiration causes hemorrhages in the lungs and chemical pneumonia.
n-heptane, normal heptane, heptanes (mix of isomers)	400	230	25	FAST	One of least toxic in class. Good substitute for hexane and other fast-drying solvents.

SOLVENT CLASS	TLV-TWA	OT	FP	ER	COMMENTS
VM & P naphtha, benzine, paint thinner	300	1–40	20–40	MED	One of the least toxic in class. Good substitute for turpentine. Odorless thinner has aromatic hydrocarbons removed.
mineral spirits, Stoddard solvent, other petroleum fractions	100	1–30	>100	SLOW	Some fractions contain significant amounts of aromatic hydrocarbons.
n-hexane, normal hexane, commercial hexanes (55% n-hexane)	50	65–250	–7	FAST	Do not use. Potent nervous system toxin causing multiple sclerosis–like disease. Extremely flammable. Substitute heptane.
hexane isomers	500	—	—	FAST	Low toxicity. Often contaminated with n-hexane.
gasoline	300	0.3	–45	FAST	Do not use. Extremely flammable. May contain skin-absorbing benzene, organic lead compounds or new, toxic anti-pollution additives.

CLASS/name, synonyms	TLV-TWA	OT	FP	ER	COMMENTS
AMIDES/AMINES					**Many are sensitizing.**
dimethyl formamide (DMF)	10	0.5–100	136	SLOW	Try to avoid. Skin absorbs.
ethanolamine	3	2–4	185	VERY SLOW	Severe skin, eye, respiratory irritant. Narcosis, liver and kidney damage reported.
diethanolamine	0.46	0.27	342	VERY SLOW	More toxic than ethanolamine. Severe skin and eye damage documented.
triethanolamine	5mg/m^3*	unk	385	**	Hazards similar to ethanolamine. Avoid. An experimental carcinogen.

*milligrams per cubic meter
**Hygroscopic: absorbs water and evaporates very slowly.

CLASS/name, synonyms	TLV-TWA	OT	FP	ER	COMMENTS
AROMATIC HYDROCARBONS					**A hazardous class; avoid if possible.**
ethyl benzene, ethyl benzol, phenyl ethane	100	0.1–0.6	59	SLOW	Eye irritation begins at the TLV.
xylene, xylol, dimethyl benzene	100	20	81–90	SLOW	Highly narcotic. Causes liver and kidney damage. Stomach pain reported.
toluene, toluol, methyl benzene, phenyl methane	50	0.2–37	40	MED	Highly narcotic. Causes liver and kidney damage.
styrene, vinyl benzene, phenyl ethylene	20	0.017–2	90	SLOW	Suspected cancer agent. Try to avoid. Skin absorbs.
diethylbenzenes: 1,3-DEB, 1,4-DEB	none	unk	~130	SLOW	Narcotic and irritating. Not well-studied.
trimethylbenzenes: 1,2,3-TMB; 1,2,4-TMB; 1,3,5-TMB	25	2.4	~130	MED	Strong narcotic and irritant. Not well-studied.
benzene, benzol	0.5	34–119	12	MED	Do not use. Causes leukemia. Skin absorbs.
CHLORINATED HYDROCARBONS					**Many in this class cause cancer. Avoid.**
1,1,1-trichloroethane, methyl chloroform	350	390	**	FAST	Causes irregular heartbeat and arrest. Cancer studies underway.
methylene chloride, dichloromethane	50	160	**	FAST	Avoid. Suspected cancer agent. Metabolizes to carbon monoxide in blood. Stresses heart.
trichloroethylene	50	82	**	MED	Suspected cancer agent. Irregular heartbeat.
'perchloroethylene, perc, tetrachloroethylene	25	47	**	MED	Suspected cancer agent. Irregular heartbeat, liver damage, skin reddens after alcohol ingestion.

CLASS/name, synonyms	TLV-TWA	OT	FP	ER	COMMENTS
chloroform	10	133–276	**	FAST	Do not use. Suspected cancer agent.
ethylene dichloride, 1,2-dichloroethane	10	6–185	56	MED	Strong intoxicant, causes liver damage, a suspected cancer agent.
carbon tetrachloride	5	140–584	**	FAST	Do not use. Cancer agent. Severe liver damage and death result when combined with alcohol. Skin absorbs.

**These solvents do not have typical flash points. They dissociate with heat or ultraviolet radiation to form toxic gases such as phosgene.

ESTERS/ACETATES					**One of least toxic classes.**
ethyl acetate	400	6.4–50	24	FAST	Least toxic in class.
methyl acetate	200	180	14	FAST	Similar to ethyl acetate.
isoamyl acetate, banana oil (100) NIC	50	0.22	64	MED	Used for respirator fit testing.

ETHERS	**Do not use. Extremely flammable. Forms explosive peroxides with air.**

GLYCOLS					**Vary greatly in toxicity.**
propylene glycol, 1,2-propanediol	50 (WEEL)	unk	210	*	Least toxic glycol. May cause allergies. Use in cat/dog food causes blood damage.
ethylene glycol, 1,2-ethandiol	C 39.4 (Ceiling)	0.1–40	232	*	Lung and eye irritant. Neurological damage and blindness at high doses.
diethylene glycol	50 (WEEL)	unk	255	*	Probably more toxic than ethylene glycol, but does not cause blindness. Skin absorbs.
triethylene glycol, triglycol	none	unk	350	*	Technically, a glycol ether, which may be a reproductive hazard—see next section.

*Hygroscopic: absorbs water and evaporates very slowly.

CLASS/name, synonyms	TLV-TWA	OT	FP	ER	COMMENTS
GLYCOL ETHERS (CELLOSOLVES) AND THEIR ACETATES					**Try to avoid, especially if planning a family.**
butyl cellosolve, 2-butoxyethanol, ethylene glycol monobutyl ether	20	0.1	141	SLOW	Affects blood, liver, kidneys. Not as toxic to reproductive system as others. Skin absorbs.
cellosolve, 2-ethoxyethanol, ethyl cellosolve, ethylene glycol monoethyl ether	5	2.7	110	SLOW	Reproductive hazard for men and women. Blood, liver, kidneys. Skin absorbs.
'methyl cellosolve, 2-methoxyethanol, ethylene glycol monomethyl ether	5	2.4	102	SLOW	Same as above.
di- and triethylene and propylene glycol ethers and their acetates	—	—	—	—	This is a large class, many of which are not well-studied. Experts suspect some harm to blood & reproductive systems.
KETONES					**Toxicity varies widely.**
acetone, 2-propanone, dimethyl ketone	500	3.6–653	–4	FAST	Least toxic. Highly flammable. Irritating to respiratory tract.
methyl ethyl ketone, MEK, 2-butanone	200	2–85	16	FAST	Causes severe nerve damage when used in combination with n-hexane.
methyl isobutyl ketone, MIBK	50	0.013	64	MED	May be more toxic when used in combination with n-hexane.
methyl butyl ketone, MBK	5	0.07–09	77	MED	Do not use. Causes permanent nerve damage.
MISCELLANEOUS					
turpentine	100	50–200	95	SLOW	Causes allergies (dermatitis, asthma), kidney and bladder damage. Use odorless paint thinner.

CLASS/name, synonyms	TLV-TWA	OT	FP	ER	COMMENTS
limonene, d-limonene, citrus oil, citrus turps, menthadiene, dipentene	30 (WEEL)	unk	unk	VERY SLOW	A pesticide, cancer drug, food additive. Acutely toxic by ingestion (rats). Inhalation causes kidney damage. Pleasant odor tempts children to drink it. More toxic than turpentine.
morpholine	20	0.011	100	SLOW	Avoid. Skin absorbs.
tetrahydrofuran	200	31	1.4	VERY FAST	Becomes explosive when oil or exposed to air. Highly narcotic.
dioxane, 1,4-dioxane	20	0.8–172	65	FAST	Carcinogen. Skin absorbs. Avoid.
cyclohexane, hexamethylene (300) NIC	200	780	1.4	FAST	Not acutely toxic. Chronic effects unknown.

12 Paints, Inks, Pigments, and Dyes

Today, almost every kind of paint, ink, pigment, and dye is used for scene and costume work. For scene painting, traditional caseins have been supplemented by a host of special theatrical paints, general market paints, fine art materials, and industrial paints. Costumers now use dyes designed for the new synthetic fabrics, fabric paints, silk screen inks, and many new materials.

WHAT ARE PAINTS AND INKS?

Paints and inks are made up of two components: a pigment and a vehicle or base. The most common vehicles are oils, water, casein, and emulsions of polymers, such as acrylics or urethanes.

Oil-Based

Traditional oil-based paint, finishes, and inks are those in which the actual coating material—usually resins or a blend of oils and resins—is dispersed in a solvent vehicle. These products must also be cleaned up with solvents. The solvents can be divided into those that damage the ozone layer and those that do not.

Only the solvents that damage the ozone layer are regulated and will be listed on paint labels as "VOC," or "volatile organic chemical" content. This means that a paint that has low VOC content may still contain a lot of solvents that are not regulated. Always get the material safety data sheet on paints, and do not rely on labels alone.

Water-Based

Water-based polymer finishes, coating materials, and inks are plastic material dispersed as tiny droplets suspended in water. The water evaporates, leaving the tiny, plastic droplets to fuse into the actual finish.

These paints are safer in general, because they can be cleaned up with water. But the products themselves usually contain some solvents dispersed in the water. Both regulated VOC solvents and unregulated ones may be found in water-based paints and inks. The class of solvent most often found in these products is the glycol ethers (see chapter 11).

The polymer resins, such as acrylic, epoxy, or urethane, do not mix with water, and they do not naturally stay in little droplets in a water emulsion. To make this polymer/water emulsion stable, large numbers of small chemical additives and preservatives are used. Examples of the types of chemicals that will be present in these types of paints are found in table 12.1.

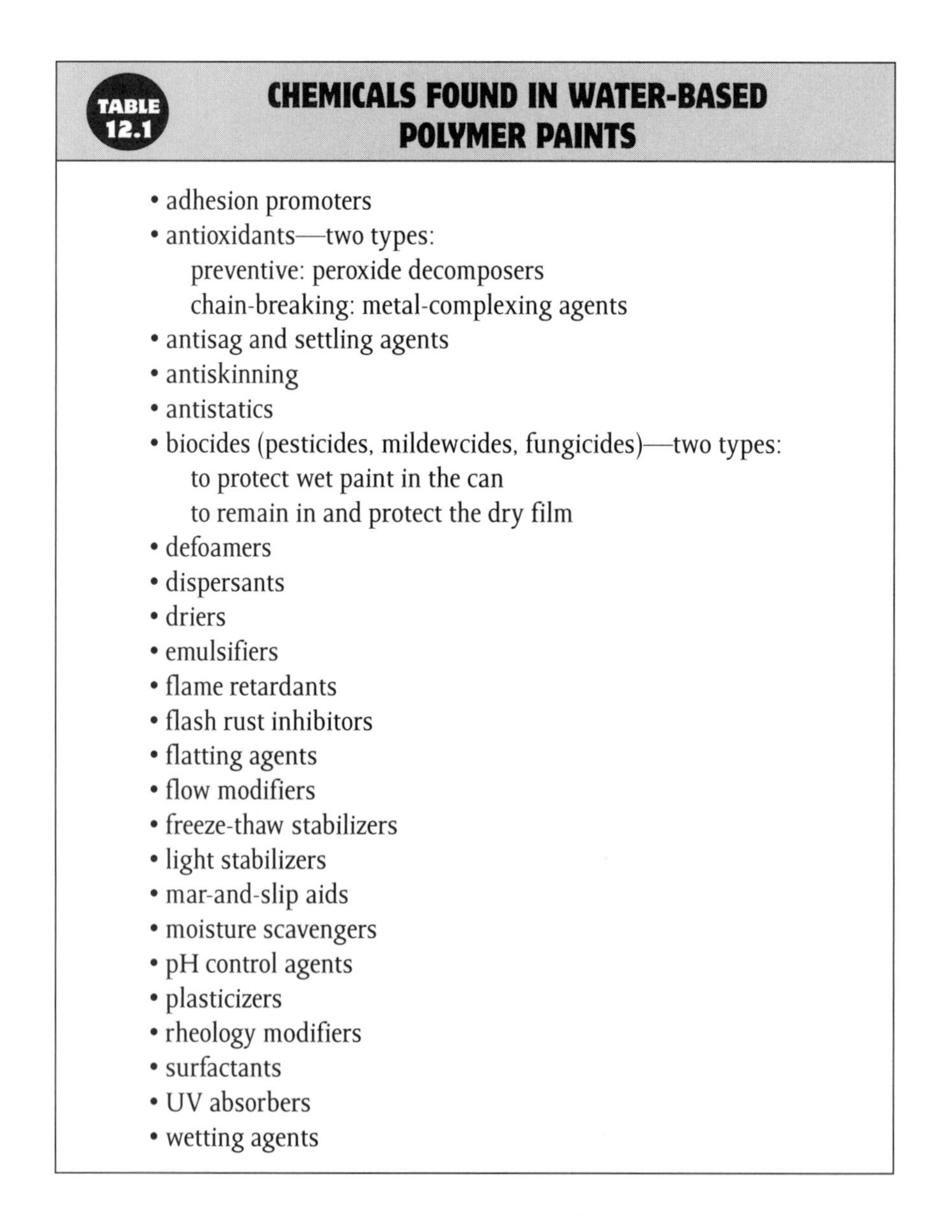

TABLE 12.1 CHEMICALS FOUND IN WATER-BASED POLYMER PAINTS

- adhesion promoters
- antioxidants—two types:
  preventive: peroxide decomposers
  chain-breaking: metal-complexing agents
- antisag and settling agents
- antiskinning
- antistatics
- biocides (pesticides, mildewcides, fungicides)—two types:
  to protect wet paint in the can
  to remain in and protect the dry film
- defoamers
- dispersants
- driers
- emulsifiers
- flame retardants
- flash rust inhibitors
- flatting agents
- flow modifiers
- freeze-thaw stabilizers
- light stabilizers
- mar-and-slip aids
- moisture scavengers
- pH control agents
- plasticizers
- rheology modifiers
- surfactants
- UV absorbers
- wetting agents

Paint additives are rarely listed on labels or in MSDSs. Their effects on people in the long term are usually unknown. They probably are responsible for some of the allergies and sensitivities some people have to paints. However, it is assumed that the most hazardous ingredients in paints are their solvents (if they contain them) and their pigments.

PIGMENTS

Pigments are divided into inorganic and organic chemical classes.

Inorganic Pigments

The hazards of these pigments have been known since Bernardo Ramazzini described the devastating illnesses associated with pigments in 1713. At this time, pigments containing lead, cadmium, mercury, and all manner of toxic metals were used.

Today, most of these same pigments are still used in art paints and inks. They are allowed to be used, because artists' materials are exempt from the paint lead laws. These laws forbid the use of lead in consumer wall paints, but exempt artist's paints, auto and boat paints, metal priming paints, and other specialty paints. Never assume any paints, except interior wall paints, are free of lead, cadmium, or other highly toxic pigments. Check labels and MSDSs before using paints.

Some inorganic pigments are not hazardous. Examples of pigments with no significant hazards include the yellow and red iron ochres, ultramarine colors, and zinc oxide whites. However, this information is not very helpful, because most of the manufacturers of theatrical paints do not identify the pigments in their products.

Organic Pigments and Dyes

Complex organic chemicals can also be used to impart color to paints, inks, and fabric dyes. The origins of organic pigments and dyes are lost in antiquity, although we know that both sprang from common natural products, such as berries, roots, minerals, and insects. When mauve, the first synthetic dye, was discovered in 1856, it catalyzed development of the whole organic chemical industry. Since then, thousands of dyes and pigments have been synthesized. Most organic colorants used today are synthetic.

When a dye is applied, it penetrates the textile in a soluble form, after which it may even react in some way that binds it to the fabric. Pigments, on the other hand, are insoluble and remain unaltered by the surface they are on. However, the colorant class of both dyes and pigments may be the same. In fact, there are some colorants that function as both dyes and pigments, because they are soluble in water (a dye) and insoluble in oil paint (pigment).

Evidence that many organic dyes and pigments are hazardous is found in the tragic history of industrial dye workers. Bladder cancer, in particular, was related to exposure to classes of chemicals called aniline or benzidine dyes.

Benzidine Pigments and Dyes

Bladder cancer, associated with this chemical, is one of the oldest-known occupational cancers. It was first associated with dyes in 1895, when a Swiss urologist noted a high incidence of bladder tumors among dye workers. In the 1970s, hundreds of cases were studied in Japan among kimono painters and dyers. And even as late as 1989, a study in the United States showed that professional artists and printmakers are at elevated risk of developing bladder cancer.[1] Also at higher risk are barbers, hairdressers, leather workers, painters, and, of course, textile workers and dyers. This does not bode well for our theatrical workers who also use dyes and pigments.

While there has never been a study of theater painters, I am troubled by what appear to be little clusters of bladder cancer that I have seen. In a network scene shop with about ten scenic employees, there are two confirmed cases of bladder cancer and two more people who are being monitored because they have blood in their urine and suspicious cell changes in their bladder walls. I also know of an art department on the campus of one midwestern university in which there are three cases of bladder cancer among the printmaking and painting faculty.

Anthraquinone Pigments and Dyes

Another large class of dyes is also being studied. So far, anthraquinone itself and five related anthraquinone dyes have all been shown to cause cancer in animals. Experts think most of these dyes cause cancer. The anthraquinone dyes are common and are found in products such as Rit® and Tintex®. Some hair dyes and common art pigments are also in this class.

One anthraquinone pigment, alizarin crimson, is almost surely a carcinogen, based on its structure. This chemical can be obtained naturally from madder root or can be made synthetically as Muralo®. In terms of toxicity, the origin of dyes or pigments is irrelevant.

BRONZING AND ALUMINUM POWDERS

Finely powdered metals of various kinds are used to metalize or glitter surfaces. These, technically, are pigments, but they have a very special hazard: most are explosive under certain conditions.

[1] "Occupational Risks of Bladder Cancer in the United States: I. White Men," *Journal of the National Cancer Institute,* Vol. 81, No. 19, Oct. 4, 1989.

Powdered aluminum, iron, magnesium, titanium, tungsten, zinc, zirconium, and alloys of these metals (e.g., bronze) are common pyrotechnic ingredients. If scenic artists, costume workers, or prop makers dust these powders onto surfaces, they need to be aware that a spark, flame, or static discharge can cause the suspended dust to burn rapidly and explosively in the air. If the dust is suspended in a container and it ignites, the sudden pressure of that burn will cause the container to explode.

These materials must be used without raising much dust. Even better, they now can be replaced with water-based emulsions or dispersions of these metal pigments. When dry, they look the same as the dry applied pigments.

Proper storage of these materials is also an interesting problem. If the shop has a pyrotechnic magazine, this would be a good place for them. They are not flammable liquids, so technically, they do not belong in a flammable storage cabinet. However, this might be another acceptable location. A separate metal cabinet with warning labels might be another solution.

IDENTIFYING PIGMENTS AND DYES

Companies selling theater paints, inks, pigments, and dyes list colors in many ways, sometimes using simple, traditional names (white, red, etc.) and sometimes fanciful names designed to attract customers. As a result, it is almost impossible to know the actual color chemicals to which these names refer.

For example, real "moly orange" is lead molybdate, a highly toxic lead pigment. Paints sold as "moly orange" today probably are not lead pigments. But it is impossible to tell what this pigment actually is, and the manufacturers of theatrical paints usually will not tell customers the real names.

This lack of proper identification means that painters and dyers with health problems can't tell their doctors precisely what chemicals they use. And if an actor reacts adversely to a dyed garment or dye-containing product, the dyer can only refer him to the specific product, not to the name of a chemical so he can avoid exposure in the future.

The inability to identify pigments and dyes also means that artists can't reliably repeat color effects. And in some cases, they cannot sell products colored with unknown chemicals in other countries. Artists selling textiles or clothing in Germany can be required by law to declare in writing that over 120 specific dyes are not used on items intended to be worn next to the skin. Other European countries are adopting this law. But U.S. and Canadian dyers usually can't comply with these new export laws, because they don't know the proper names of the dyes they used and whether or not the banned dyes are on their products.

Color Index

The answer to this identification problem is simple: As an industry, we need to demand that suppliers identify their colorants with the internationally accepted system of identification used in the Colour Index (C.I.).[2] This is a nine-volume index published jointly by the Society of Dyers and Colourists in the United Kingdom and the Association of Textile Chemists and Colorists in the United States. The C.I. classifies and provides technical information on all colored materials, including all classes of dyes and pigments.

RULES FOR USING PAINTS, INKS, AND DYES

- Obtain Material Safety Data Sheets (MSDSs) on all paint and dye products. If pigments and dyes are not identified by their C.I. names or numbers, ask your supplier for this information. Purchase from suppliers who provide this information whenever possible.
- Treat all colorant products as highly toxic, since they probably are.
- Use MSDSs and product labels to identify the hazards of any toxic solvents, acids, or other chemicals in dyes, paints, inks, mordants, or other materials.
- Use water-based products whenever possible to avoid solvent cleanup. Check MSDSs to be sure water-based materials do not themselves contain significant amounts of solvents.
- Buy premixed paints and dyes if possible. Dyes packaged in packets that dissolve when dropped unopened into hot water also can be handled safely. Pigments and dyes are most hazardous and respirable in a dry, powdered state.
- Weigh or mix dye powders or other toxic powders where local exhaust ventilation is available, or use a glove box (see figure 12.2).
- Avoid dusty procedures. Sanding dry paints, sprinkling dry pigments or dyes on wet paint or glue, and other techniques that raise dust should be discontinued or performed in a local exhaust environment. If you use such processes without ventilation, choose pigments (do not use dyes) known to be of low toxicity, wear a respirator, and use wet-mop cleaning methods.
- Choose brushing and dipping techniques over spray methods whenever possible.

---

[2]*Colour Index International,* 9 volumes, 1971–1992, Society of Dyers and Colourists and American Association of Textile Chemists and Colorists. Contact: P.O. Box 12215, Research Triangle Park, NC 27709-2215, (919) 549-8141, *www.aatcc.org.*

FIGURE 12.2

- Spray paints or dyes only under local exhaust conditions, such as in a spray booth. A proper respirator (dust/mist respirator for water-based paints; paint, lacquer, and enamel mist respirator for solvent-containing products) may provide additional protection.
- Avoid skin contact with paints and pigments by wearing gloves or using barrier creams. Use gloves with dyes. Wash off paint splashes with safe cleaners like baby oil followed by soap and water; nonirritating, waterless hand cleaners; or plain soap and water. Never use solvents or bleaches to remove splashes from your skin.
- Wear protective clothing, including a full-length smock or coveralls. Leave these garments in your studio to avoid bringing dusts home. Wear goggles if you use caustic dyes or corrosive chemicals.
- Work on easy-to-clean surfaces, and wipe up spills immediately. Wet-mop and sponge floors and surfaces. Do not sweep.
- Avoid ingestion of materials by eating, smoking, or drinking elsewhere than your workplace. Never point brushes with your lips or hold brush handles in your teeth. Never use cooking utensils for dyeing. A pot that seems clean can be porous enough to hold hazardous amounts of residual dye. Wash your hands before eating or smoking.
- Keep containers of paint, powdered dyes and pigments, solvents, etc., closed when you are not using them.
- Follow all solvent safety rules if you use solvent-containing products.
- Arrange for regular blood lead tests if you use lead-containing paints or pigments.

13 Plastics and Adhesives

Plastics—which have changed our daily lives greatly in the last few decades—have done some revolutionizing in theater as well. Theater workers use plastics as casting materials, foams, glues, adhesives, structural elements, artificial snow, costume padding, textiles, gels, and much more.

WHAT IS PLASTIC?

A plastic, or "polymer," is created when a chemical called a monomer reacts with other molecules like itself to form large molecules, often in long chains. This reaction is called polymerization. For example, when a monomer called methyl methacrylate is polymerized, it becomes *poly*methyl methacrylate, better known as Lucite or Plexiglas. Mother Nature makes some monomers and polymers as well. Natural rubber and linseed oil are examples.

Some polymers are capable of a second reaction in which the long chains are linked together laterally (side by side). This reaction is called cross-linking. For example, liquid polyester resin becomes a solid material when it is reacted with a cross-linking agent like styrene.

Long-chain and cross-linked polymers react differently when exposed to heat. Heat usually will deform or mold long-chain polymers into new shapes. These polymers are called "thermoplastics." On the other hand, heat will not deform cross-linked polymers, and these are called "thermoset" plastics.

Chemicals that can cause monomers and resins to react have many trade names, including activators, actuators, catalysts, curing agents, hardeners, or initiators. In this book, the term "initiator" will be used in most cases.

Special attention should always be paid to the initiator or hardener used with any resin system, because most initiations are highly hazardous and reactive.

HAZARDS OF PLASTIC RESIN SYSTEMS

Plastics are most hazardous during polymerization and cross-linking. Most monomers, initiators, and cross-linkers are very toxic. The hazards vary with each type of plastic resin system. Be sure you know which plastic you are polymerizing.

During polymerizing, it is important to follow product directions precisely. If directions are followed, the reaction should bind the hazardous chemicals into the solid plastic. Exceptions occur when the reaction is not complete because mixing was not uniform, the proportions were not correct, or some other factor. In these cases, unreacted monomer or other chemicals may be left to off-gas indefinitely or to render the plastic's dust toxic when it is machined.

POLYESTER RESIN SYSTEMS

Hazardous chemicals used in these systems include: the cross-linking agent, which is usually styrene; ketone solvents such as acetone, used to dilute the resin or for cleanup; the initiator, which is an organic peroxide such as methyl ethyl ketone peroxide; and fiberglass, sometimes used for reinforcement.

Styrene is a highly toxic, aromatic hydrocarbon solvent that can cause narcosis, respiratory-system irritation, liver and nerve damage, and is a suspected carcinogen. Acetone is a less toxic solvent, but is extremely flammable.

Fiberglass dust can cause skin and respiratory irritation. There are many other compounds in polyester resin systems that initiate, promote, or accelerate the reaction. The hazards of many of these chemicals are not well studied.

Organic peroxides, such as methyl ethyl ketone peroxide and dicomarol peroxide, may be used to initiate the reaction. These are among the most hazardous chemicals found in the shop. Methyl ethyl ketone peroxide has caused blindness when splashed in the eyes, can form an explosive mixture with acetone, and converts to a shock-sensitive explosive material after a time. (See Organic Peroxide Initiators for a more complete discussion of hazards.)

Organic Peroxide Initiators

These initiators are used to cross-link or initiate a number of types of plastic resin systems. Examples are methyl ethyl ketone peroxide, dicomarol peroxide, and benzoyl peroxide. Vapors of these chemicals may cause eye and respiratory irritation. Liquid organic peroxides splashed in the eye have caused blindness. However, primary hazards are safety and fire hazards.

Most organic peroxides can become shock-sensitive and explosive when they get old, oxidized, and unstable. They can also create highly unstable mixtures with flammable materials such as the acetone used to clean up resin.

Organic peroxides burn vigorously. Once on fire, they can provide their own oxygen to the flame and cannot be extinguished by removing oxygen. For example, if organic peroxides ignite after being spilled on clothing, the fire cannot be extinguished and must burn until spent.

Because of organic peroxides' fire and explosion hazards, they are usually sold mixed with inhibitors. Even so, these mixtures have been known to burn quietly until all the inhibitor is burned off—then, the fire suddenly intensifies.

Other Ingredients

Polyester resin systems also often contain a few percent of very toxic organic metal compounds, such as dibutyl tin dilaurate, or organophosphate chemicals, such as triethyl phosphate. These chemicals probably absorb through the skin and function in the body similar to pesticides.

Precautions for Using Peroxide Initiators

- Obtain Material Safety Data Sheets and product information on peroxides. Pay special attention to information about reactivity, fire hazards, and spill procedures.
- Keep date of purchase and date of opening on the container label. Dispose of peroxides after a year if unopened or after six months once opened. Never purchase peroxides if containers are damaged or irregular.
- Store peroxides separately from other combustibles and flammables. Always keep peroxides in their original containers.
- Do not store large amounts of peroxides without consulting fire laws.
- Do not heat peroxides or store them in warm areas or direct sunlight.
- Never dilute peroxides with other materials or add them to accelerators or solvents.
- Wear protective goggles when containers are open or when pouring peroxides.
- When mixing small amounts of resins and peroxides, use disposable containers. Soak all tools and containers in water before disposing of them.
- Clean up spills immediately, in accordance with Material Safety Data Sheet directions. Inert materials such as unmilled fire clay are usually recommended to soak them up. Clean up peroxide-soaked material with nonsparking, nonmetallic tools. Do not sweep them; fires have started from the friction of sweeping itself.
- If peroxide spills on your clothing, remove your clothes immediately and launder them separately and well before wearing them again.

- When discarding unused peroxide or fire clay/peroxide mixtures, first react them with a 10 percent sodium hydroxide solution to prevent fires. Alternatively, unused peroxide can be reacted with resin and the solid plastic discarded safely.

URETHANE RESIN SYSTEMS

Two-component or A/B urethane foam, casting, and coating resins have become widely used in scenic work. They can be purchased in amounts ranging from fifty-five-gallon drums to small double-nozzle spray cans. They are among the most dangerous products we use. This author knows two scenic artists who were partially disabled and another who is now unable to work at all after working with urethanes.

It is not the resin in these products that is most hazardous. Actually, there is no "urethane resin." Instead, the resin half of the product can be many different resins, such as polyesters, polyethers, polyols, epoxies, and so on. These resins do not become polyurethane until they are reacted with a diisocyanate initiator. While the resin may contain solvents or other hazardous ingredients, it is the other half of the product, the diisocyanate hardener, that is deadly.

Diisocyanate hardeners are capable of causing severe respiratory allergies and lung damage. Most notably, they cause an incurable and progressive occupational illness called isocyanate asthma. Not all workers, of course, develop this disease. But among those that do, the pattern typically is that they are exposed a number of times without incident and then suddenly begin to develop respiratory reactions. Other allergic reactions are also related to isocyanate exposure, including anaphylactic shock, which has resulted in deaths of workers using urethanes.

To protect people from these effects, extremely restrictive workplace air-quality limits have been set (see table 13.1 below). And since there is no respirator cartridge approved for isocyanates, local ventilation or air-supplied respiratory protection systems must be employed. Leave freshly cast or foamed objects in the exhaust system for several hours or overnight until they have completely set and finished off-gassing before taking them into the general workspace.

Some of the MSDSs and product labels will provide stern warnings about the dangers of the diisocyanates. Others will not. The products *without* strong warnings usually contain one of the forty or more isocyanates for which no standards exist. This does not mean that these diisocyanates are safe. It means they have not been tested. Yet, most experts and even the EPA believe ". . . that it is reasonable to anticipate that all members of the diisocyanate category will exhibit chronic pulmonary toxicity. . . ."[1]

---

[1]59 FR 61454, November 30, 1994, EPA: Addition of Certain Chemicals; Toxic Chemical Release Reporting: Community Right-to-Know; Final Rule.

Do not accept claims that a product containing an isocyanate is safe because it is prepolymerized or modified without checking out these claims with experts who are not trying to sell you a product.

NATURAL RUBBER

Rubber can be considered to be a plastic resin manufactured by nature. Natural rubber is derived from sap drawn from hevea trees. It contains a chemical called isoprene, which can be reacted to form a polymer or plastic-like material called *poly*isoprene—the fancy name for natural rubber.

Natural rubber can be found in thousands of products. Some of these include surgical and chemical gloves, condoms, balloons, rubber cement, many latex molding products, surgical and eyelash adhesives, and many special-effects latex makeup products. Almost all rubber tires contain some natural rubber, although the main ingredient is more likely to be a synthetic rubber. The synthetic rubbers do not contain the sensitizing natural proteins.

Water-based natural rubber latex systems are commonly used to make molds. Solvent-containing rubber products are also used. In addition, rubber cement and some contact cements are rubber dissolved in solvents. These products usually contain very toxic solvents, such as hexane, which is especially toxic to the nervous system (see table 13.1). Choose rubber products that replace hexane with less toxic heptane.

Hazards of Rubber

Rubber itself is a hazard due to its well-documented ability to cause allergies. Symptoms include skin rash and inflammation, hives, respiratory irritation, asthma, and systemic anaphylactic shock. Between 1988 and 1992, the FDA received reports of one thousand systemic shock reactions to natural rubber gloves and medical devices. By June 1996, twenty-eight latex-related deaths had been reported to the FDA. Now, actions have been taken to avert these tragedies.

The Cause

Natural latex contains many impurities, including about 200 different proteins made by the rubber tree. Fifty of these proteins can cause allergies. The proteins cannot be completely removed from finished natural rubber products. Gloves labeled "reduced proteins" are less sensitizing, but all-natural rubber products can cause allergy. For this reason, the FDA now prohibits the word "hypoallergenic" on latex glove labels.

What Is Latex?

The word "latex" simply means any plastic (polymeric) substance in an essentially water medium. For example, acrylic latex paint is a water dispersion of acrylic polymer. Acrylic latex paint contains no rubber. In fact, latex paints in general almost never contain natural rubber.

Who Reacts?

The best data on allergy comes from medical professions, since they use gloves every day. Various studies estimate that somewhere between 10 and 17 percent of health care workers are now allergic to rubber. This means that better than one out of every ten people who frequently use rubber products will develop the allergy.

Precautions to Use with Rubber Products

- Try to switch to synthetic plastics, such as nitrile plastic gloves or synthetic rubber casting materials.
- If rubber latex gloves must be used, use reduced-protein gloves and powder-free gloves. The powder can increase exposure through skin contact and inhalation.
- Reduce other daily exposures. Use nonlatex gloves for food preparation and housekeeping.
- If you are a person who tends to get allergies, be alert for reactions so you can take action promptly.
- Wash your hands with a mild soap immediately after removing latex gloves or working with rubber-containing products.
- Never use hand lotion or barrier creams under gloves or with casting products, since they can leach the proteins out of the rubber and make the situation worse. Lotions and creams can also make rubber and plastic gloves more easily penetrated by germs and chemicals.
- Clean areas contaminated with latex-containing dust or spilled latex products without raising dust or splashing wash water on yourself.
- See your doctor if allergy symptoms start. There is no cure for latex allergy, but some medications can reduce symptoms. If you are diagnosed with a serious rubber allergy, avoid all products containing natural rubber.
- Learn about your latex allergies. For instance, you should know that allergies to certain foods, like avocados, potatoes, bananas, tomatoes, chestnuts, kiwi fruit, and papaya, are also associated with latex allergy.

SILICONE MOLD-MAKING MATERIALS

Two types of silicone resin systems are commonly used to make molds. The first is a single-component system that cures by absorbing atmospheric moisture. The second is a two-component system that cures by means of a peroxide (see section above, Organic Peroxide Initiators). Both systems contain solvents, such as acetone or methylene chloride. Some contain small amounts of rather toxic organic metal compounds, such as dibutyl tin laureate.

Single-component systems may release acetic acid or methanol into the air. Acetic acid vapors are irritating to the eyes and respiratory tract. Methanol is a nervous system poison. Two-component systems often contain chemicals that can damage the skin. Some also contain methylene chloride, which can cause narcosis and stress the heart (see table 13.1 below).

EPOXY RESIN SYSTEMS

Epoxies are used for paint and ink vehicles, casting, laminating, and molding. They are also common adhesives and putties. Most are two-component systems. After they are mixed, the resulting epoxy gives off heat, which vaporizes any solvents in it. An excess of hardener can cause the epoxy to heat to the point of decomposition and ignition.

Epoxy resins can irritate the skin. They may also contain varying amounts of solvents. Common solvents in epoxy include the glycidyl ethers, which have caused reproductive and blood diseases in animals, including atrophy of testicles, damage to bone marrow, and birth defects.

Epoxy hardeners are toxic and highly sensitizing to the skin and respiratory system. Almost 50 percent of industrial workers regularly exposed to epoxy develop allergies to them.

Some products contain initiators that are far more toxic than others. These products should be avoided. For example, some epoxy systems employ initiators in a class of highly toxic chemicals called aromatic amines. Some specialized epoxies use peroxide initiators (see hazards above). Other epoxies use much safer initiators called aliphatic amines, or polyamides. Look on MSDSs for these classes of chemicals, and choose the safer products (or call and ask ACTS to help you evaluate the MSDS).

OTHER RESIN SYSTEMS

There are a number of other resin systems, such as those employing methyl methacrylate (MMA) alone or occasionally in combination with other monomers. Some of these systems need special precautions, because they involve elevated pressures and/or temperatures.

New plastics and products are being formulated regularly. The hazards of some of these new products are not well-known, and some manufacturers consider the ingredients trade secrets.

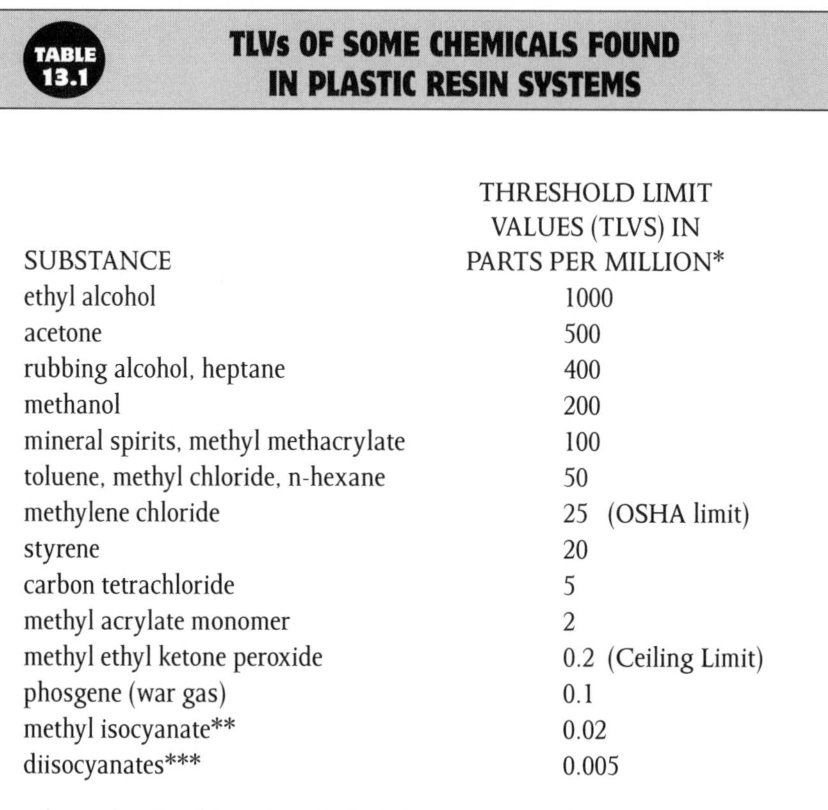

TABLE 13.1

TLVs OF SOME CHEMICALS FOUND IN PLASTIC RESIN SYSTEMS

SUBSTANCE	THRESHOLD LIMIT VALUES (TLVS) IN PARTS PER MILLION*
ethyl alcohol	1000
acetone	500
rubbing alcohol, heptane	400
methanol	200
mineral spirits, methyl methacrylate	100
toluene, methyl chloride, n-hexane	50
methylene chloride	25 (OSHA limit)
styrene	20
carbon tetrachloride	5
methyl acrylate monomer	2
methyl ethyl ketone peroxide	0.2 (Ceiling Limit)
phosgene (war gas)	0.1
methyl isocyanate**	0.02
diisocyanates***	0.005

*An explanation of these air-quality limits is found in chapter 8.
**This is the chemical that killed thousands of people in an accident in Bhopal, India.
***Includes TDI (toluene diisocyanate), MDI (methylene bisphenyl isocyanate), hexamethylene diisocyanate, isophorone diisocyanate, methylene bis (4-cyclohexylisocyanate).

GENERAL PRECAUTIONS FOR TWO-COMPONENT RESIN SYSTEMS

- Never use resin systems without reading the MSDSs first. The hazards vary, and you may need to prepare in advance for safety equipment.
- Most resin systems should be used with ventilation. Polyesters, urethanes, and most solvent-dissolved resins will need local exhaust ventilation. Silicon and natural rubber latex products only need a small amount of dilution ventilation.
- If backup respiratory protection is needed, consult the MSDS to choose the right type. Two-component urethanes will require air-supplied respiratory protection systems, since there are no cartridges

approved for isocyanates. Organic vapor cartridges can be used to capture styrene from polyester resins. Add a dust prefilter to the respirator if you use fiberglass or if you sand the finished plastic.

- Wear gloves and chemical splash goggles when handling and pouring materials. Protection from some plastic resin chemicals requires special types of gloves. Consult the glove manufacturer's permeation data (see chapter on personal protective equipment).
- Wear clothing that covers your arms and legs. Remove clothes immediately if they are splashed with resins or peroxides. Always remove clothing completely after work, then take a shower.
- Handle peroxides correctly by following the advice in the section on organic peroxide initiators in this chapter. Be especially careful to avoid splashes in the eyes, and never mix peroxides with solvents, such as acetone.
- Follow all precautions for using solvents, such as cleaning up spills immediately, disposing of rags in approved, self-closing waste cans, and the like (see chapter 11, Solvents).
- Try to use disposable containers for mixing resins so that it will not be necessary to clean them with strong solvents. Small amounts of resins can be mixed in paper cups with wooden sticks. If you need reusable containers, use polyethylene or stainless steel containers.

FINISHED PLASTICS

Rather than working with resin systems, it is safer to use sheets, precast foam and boards, films, beads, or blocks of finished plastic. Even so, there are emissions and dusts created during use that must be considered.

Heating and Burning

When plastics are cut or heated, decomposition products are released, and these products can be hazardous. Processes during which this can occur include sawing, sanding, hot knife or wire cutting, press molding, drilling, grinding, heat shrinking, vacuum forming, extruding, burnout casting, torching, and melting.

The chemicals produced during heating or burning of plastics under normal shop conditions are very complex. Literally hundreds of chemicals are generated, some of which invariably will be very toxic. The complex chemical content of the emissions means that no respirator will be successful in capturing all of these chemicals. Heat decomposition emissions should be controlled by ventilation.

MSDSs for plastics are often misleading with respect to heat-generated emissions. Each MSDS has a section on hazardous decomposition products.

However, the data in this section almost always relates to high-heat incineration in the presence of oxygen. In this case, a styrene plastic, for example, would be reduced to carbon dioxide and water. This is not what is released when plastics are wire-cut or power-sawed.

The MSDS decomposition data is useful to identify the type of plastic. For example, if you see hydrochloric acid or chlorine listed, the plastic contains chlorine and is probably a polyvinyl chloride (PVC). If cyanide is a decomposition product, the plastic must be a nitrogen-containing plastic, such as polyurethane, melamine, urea formaldehyde, or nylon.

Dusts

The dusts of some plastics are sensitizing and can cause skin, eye, and respiratory symptoms in some people. This is more likely to occur in plastics whose resins and initiators are also sensitizers, such as urethanes and epoxies.

Allergies may also be due to the many additives plastics contain. Typical additives include plasticizers (used to achieve the desired softness), stabilizers, colorants (dyes and pigments), fillers, fire retardants, inhibitors, accelerators, and more. Some of the common plasticizers (some of the phthalate esters) are known to cause cancer in animals. However, the vast majority of these additives' hazards are unknown.

Plastic adhesives also contain toxic solvents that require precautions (see chapter 11).

General Precautions for Finished Plastics

- Use good dilution ventilation or local exhaust ventilation. Use water-cooled or air-cooled tools, if possible, to keep decomposition to a minimum. When heat-forming plastics, use the lowest possible temperature.
- Add a vacuum attachment to sanders, saws, and other electric tools to collect dust.
- Try to provide ventilation, since no respirator cartridge will capture all the emissions from decomposing plastics. To provide even partial protection, a respirator would have to have a cartridge rated for particulates from sanding or sawing; organic vapors; acid gasses for hydrochloric acid gas from decomposing polyvinyl chloride plastics; and formaldehyde from release of aldehydes on heating.
- Clean up all dust carefully by wet mopping. Do not sweep.

ADHESIVES AND GLUES

Most glues and adhesives are either synthetic or natural polymers. Some are single-component systems, like white glue. Some are two-component systems, like epoxies. Most are less hazardous than casting plastics, simply because smaller amounts are used.

COMPOSITION AND HAZARDS OF COMMON ADHESIVES

Airplane glue: Plastic dissolved in solvent, usually acetone. No hazards except solvents.

Cyanoacrylate, Super Glue®, Krazy Glue®: Plastic monomer that cures on exposure to air. Potent eye irritant. Fast curing time can cause unplanned adhesion of body parts.

Elmer's®, white glue, PVA: Polyvinylacetate plastic in a water-based emulsion. One of the least hazardous glues, unless poorly manufactured (these contain unreacted monomer).

Epoxies: See above.

Glue sticks: Plastic and/or mucilage materials. No known significant hazards.

Hot glue guns: The odor created during use is primarily from decomposition and from plasticizers, such as the phthalate esters. Provide some ventilation.

Rubber cement: Rubber dissolved in solvent, often n-hexane or n-heptane. N-heptane is a safer solvent. Some people are allergic to the rubber.

Silicone adhesives: See above.

Urea formaldehyde and **phenol formaldehyde:** Resin dissolved in solvents or water-based materials. Also a plywood and pressboard adhesive. These off-gas formaldehyde. Avoid if possible.

Wallpaper paste: Usually wheat or methyl cellulose and preservatives. May contain large amounts of pesticides and fungicides. Use wheat or methyl cellulose paste made for children, which should be free of pesticides and have no significant hazards.

White paste, library paste: No significant hazards. The preservative is often wintergreen, which sometimes tempts children to eat the paste.

14 The Shops

Scene shops, electric and lighting shops, prop shops, and costume shops all have many hazards in common that can be addressed by following general safety rules.

GENERAL SHOP SAFETY RULES

- Wear special work clothes, and remove them after work. If possible, leave them in the workshop, and wash them frequently and separately from other clothing—especially separately from children's clothing. If the workplace is dusty, wear some form of hair covering (hair is a good dust collector). Do not wear loose clothing, scarves or ties, or jewelry. Tie back long hair.
- Do not eat, smoke, or drink in shops or other environments where there are toxic materials. Dust, after all, settles in coffee cups. Sandwiches can absorb vapors. And hands can transfer substances to food and cigarettes for subsequent ingestion or inhalation. In addition, when substances are inhaled through a cigarette, the cigarette's heat can convert some materials into more hazardous forms.
- Wash hands carefully after work and each time before eating, applying makeup, or taking comfort breaks. Wash hands after removing chemical-resistant gloves or natural rubber latex gloves.
- Only do processes that the shop is equipped to do safely. For example, if there is no spray booth, don't used spray products.
- Make sure all containers, even those into which materials are transferred for storage, are labeled clearly, noting both their contents and hazards. Keep labels free of paint drips or any kind of damage that obscures the printing.

- Use unbreakable containers whenever possible.
- Keep all containers closed except when using them, in order to prevent escape of dust or vapors or contamination of the products.
- In case of accidental skin contact with chemicals, wash affected area with lots of water, and remove contaminated clothing. Check the product's label and/or MSDS for additional first-aid advice.
- If chemicals are splashed in your eyes, flush your eyes in the emergency eyewash for at least fifteen minutes, and get medical advice. Methods for holding your lids open during flushing, how to summon help, proper maintenance and use of the type of eyewash in your area, and related subjects should be covered in your shop's safety training.
- Do not use any cleaning methods that raise dust. Wet-mop floors or sponge surfaces. Highly toxic dusts can be cleaned up with a HEPA-filtered vacuum. Do not use an ordinary shop vacuum to clean up toxic or very small particulate dusts.
- Organize storage wisely. For example, store heavy materials near the floor so they cannot fall, and do not store anything too high to reach easily, unless there is a suitable ladder in the room.
- Store chemicals that react with each other separately. Examples include resins and their curing agents, bleach and ammonia, and photo developers and fixers. Check each product's MSDS for advice.
- Store flammable liquids, spray can products, and other flammable materials in an approved flammable-storage cabinet. (See chapter 11 for additional rules about flammable liquids.)
- Do not store materials that are flammable, combustible, or explosive (e.g., aluminum powders) near exits or entrances. Keep sources of sparks, flames, UV light, and heat (as well as cigarettes) away from flammable or combustible materials.
- Clean up spills immediately. Prepare for spills by stocking chemical adsorbents or other materials to collect spills, self-closing waste cans, and respiratory protection, if needed. If a serious spill occurs, call the fire department's emergency (HAZMAT) responders. A better strategy would be to purchase and store chemicals only in quantities that would not constitute a significant spill.
- Dispose of waste or unwanted materials safely and in accordance with label advice and MSDSs. When in doubt, contact your local department of environmental protection to find out how to dispose of specific materials. Engage a waste disposal service for removing waste. Do

not pour solvents, paints, or any regulated pollutants down drains. Nonpolluting aqueous biodegradable liquids may be poured down the sink one at a time, with lots of water.

- If fire extinguishers are part of your facility's fire protection system, make sure your employer is holding annual training on how to use them. Workers should be trained the first day on the job about use of the fire extinguishers, the fire suppression system, emergency procedures, and evacuation.
- Practice good housekeeping. Keep the workplace clean and organized. Never let projects or debris block aisles or become trip hazards. Empty waste cans daily.

SPECIAL SHOP PRECAUTIONS

The precautions needed to keep workers safe in the carpenter shop are covered in chapter 16. The hazards of the welding shop are covered in chapter 17. Some special precautions for other shops are covered here.

Prop Shops

Building, remaking, buying and renting props sometimes involves working with strange and sundry materials. Prop workers need a broad knowledge of old and new materials. A few of the very odd facts that have turned up when prop people have had questions about special materials are these:

- Beware of thrift shop purchases. Each year, the U.S. Consumer Product Safety Commission (CPSC) recalls 250 to 300 hazardous products. Not all these products are sent back to the manufacturer, and CPSC wonders where they went. The recalled products were found in thrift stores and being sold online! Included were hair dryers that could electrocute users, clothing that did not meet flammable fabrics standards, electric lamps and candles that could start fires, and much more. A checklist of recalled products is on the CPSC Web site at *www.cpsc.gov*.
- The backs of antique mirrors were often silvered with a mercury material that reconverts to tiny beads of liquid mercury. The liquid mercury outgases mercury vapor to the air, where it can be breathed. The mercury also absorbs through the skin. Avoid using old mirrors that have tiny, shiny dots of mercury on their backs.
- Some woods are so sensitizing that dermatitis has resulted from using wooden-handled tools. These woods include rosewood, iroko, and ebony. Handle with care.

- Old stuffed birds, animals, moose heads, and the like often shed considerable amounts of arsenic trioxide powder, which was used by taxidermists to preserve the skins. Some are also filled inside with asbestos. Newer taxidermied animals usually are not this hazardous.
- Fiestaware, especially the orange ware, and some types of depression glassware are radioactive. Using Fiestaware for food will expose actors to both uranium and lead. Even handling the ware can expose you.[1,2] Lead crystal also releases lead into food.
- Some of the brightly colored wires used for telephone and electrical transmission are lead pigmented and have even caused lead poisoning in telephone installers who cut and pulled off the plastic ends with their teeth. If you heat these wires to the point that the plastic darkens, the lead will also fume into the air.
- Old and water-damaged materials can grow molds. Some of these molds are actually toxic (see chapter 15).

Always check into the hazards of antique and secondhand materials. You will find out some amazing things.

Electrical/Lighting Shops

These shops have a very special problem if the work involves soldering electrical connections. Soldering more than thirty days per year in any facility triggers the OSHA Lead Standards, requiring personal monitoring of the person(s) doing the job, training, blood tests, and more.

The best strategy is to switch to the lead-free solders that are now on the market. Most of them are a little harder to use than the old 60/40 and 50/50 tin and lead solders. Nevertheless, more and more industries are using them, since they do not require the expensive ventilation and record keeping required under OSHA rules. Some of the new solders are alloys of silver, which are a bit more expensive, but they also conduct electricity better than lead.

Another strategy is to do the tests for lead that OSHA requires under conditions that will prove to OSHA that very low exposures are occurring. If these tests turn out to show that lead exposure is well below OSHA's action limit,

[1]"Accidental Contamination From Uranium Compounds Through Contact With Ceramic Dinnerware," R. W. Sheets and C. C. Thompson, *Science of the Total Environment*, 175, (1995) 81–84.

[2]"Release of Uranium and Emission of Radiation from Uranium-Glazed Dinnerware," R. W. Sheets and Sandra L. Turpen, *Journal of Radioanalytical and Nuclear Chemistry*, Vol. 235, #1-2, (1998) 167–171.

then only a few of the Lead Standard (1910.1025) precautions are needed. This can be done by setting up ventilated soldering stations.

The ventilation can consist of a special self-contained HEPA-filtered fume collection system, which takes up little space and costs well under $1,000. These are routinely used in the electronics industry for small soldering jobs. One good catalog supplier is:

Nederman
39115 West Warren Avenue
Westland, Michigan 48185
(313) 729-3344

Film Processing/Cleaning Labs

Toxic solvents are commonly used for film cleaning (see chapter 11). Special ventilation is needed for use of film cleaners. If photochemical processing is done, a whole other set of hazards is present. I suggest these workers obtain a copy of *OvereXposure: Health Hazards in Photography,* Susan Shaw and Monona Rossol, 2nd Edition, Allworth Press, 1991.

COSTUME AND WARDROBE SHOPS

A special section on costume and wardrobe workers is in order, because the hazards of these workers are underappreciated in our industry. They also are commonly not paid the same wage as other craftspeople. Parity must be established for these workers, and the hazards they face must be addressed.

The Hazards

In 1996, there was a crackdown on operators of garment industry sweatshops in California. Inspectors noted the following common violations:[3]

- Long hours
- Exposed wires, spliced cords, live parts, grounding problems
- Lack of OSHA injury- and illness-prevention programs
- Belt guards missing on sewing machines
- Blocked aisles
- Locked fire doors
- No bloodborne pathogen program (for needle punctures)
- Sanitation issues

[3]Bureau of National Affairs: Occupational Safety and Health Reporter, 25(49), May 15, 1996, p. 1704.

In my experience, it is a rare costume shop or wardrobe room that does not have at least one of these hazards.

In addition to garment cutting, costume workers use fabric dyes, shoe dye sprays, paints, adhesives, and other toxic products. Wardrobe workers use solvent spot removers and cleaners, detergents, and bleaches. These workers need the same hazard communication training as other workers. They should be familiar with all the material in this book, and in particular as it applies to solvents (chapter 11), dyes and pigments (chapter 12), and plastics and adhesives (especially adhesives; chapter 13).

These workers also are routinely given old, soiled, and even moldy materials to clean and repair (see Chapter 14). They need training about biological hazards, such as bloodborne pathogens and molds.

Bloodborne Pathogens

This standard (1910.1030) protects workers from exposure to blood and other body fluids. Three of its provisions apply to costume work:

- It requires that sharp items (such as needles) that are contaminated with blood or other body fluids be disposed of in a medical biological hazards (sharps) container
- Significant amounts of bloody materials, such as bandages or packing from toe shoes, must be placed in a medical infectious waste container
- Workers must be formally trained about the hazards and procedures for minimizing pathogenic hazards, including training about the use of gloves and disinfectants

Secondhand Costume Sources

Thrift shops are sources of recalled and illegal products (see Prop Shops section above). It is common to find garments there that do not meet flammable fabrics laws and children's garments with drawstrings that can strangle and buttons that are a choking hazard. Check out secondhand clothing by looking for recalled products at *www.cpsc.gov*. For productions in which fire is used on stage, do not assume any costume material is flame-retardant. Test fabrics before and after washing.

Special Dye Hazards

Costumers often use hot dye baths, which release small amounts of dye impurities and other substances into the air. More materials are released if fabrics are bleached or stripped. Bleaches release highly corrosive gases. Some of the dye strippers are carcinogens (e.g., thiourea). And when the dye color in the fabric fades or disappears, it means that the dye chemicals have been broken down. They usually revert to their more toxic precursors, such as aniline, benzidine, or anthraquinone. These highly toxic chemicals can also become airborne in small amounts.

Special ventilation should be provided for dye baths. A good, inexpensive system is one originally designed for ceramic kilns, but it works well for round-dye baths. The system has a circular hood that can be raised or lowered over the bath, and the exhaust only requires an opening to the outside to accommodate a flexible drier hose duct. Details can be obtained by writing to:

Richard Smith
Vent-A-Kiln Corporation
621 Hertel Avenue
Buffalo, New York 14207
(716) 876-2023

Fiber Hazards

It is now clear that all fibers, large and small, natural or synthetic, can cause lung diseases. Fibrous materials that now have been associated with lung diseases include cotton, jute, hemp, and many other natural fibers. Synthetic nylon flocking fibers are now known to cause a disease called flock workers disease. Recently, polyethylene fibers used to simulate snow also were associated with lung problems (see chapter 20).

It's really a no-brainer: We were designed to breathe air, not fibers. Keep dusts from fabrics cleaned up, and provide exhaust vents for dryers.

15 Mold and Other Biohazards

Almost every building material and most of the products we use at work can be food for microorganisms. Molds, bacteria, viruses, and other microorganisms are ever-present to invade our space. We need to know more about these hazards in order to control them.

MOLD: NOTHING TO SNEEZE AT

You've seen it, you've smelled it: It hides in dank storage areas, lurks on costumes, grows like fur on damp scenery, and crawls up the walls of old buildings. Mold is not just an annoyance. It can be deadly.

But how much mold is too much? We don't want to become paranoid about a few black lines between the tiles in the bathroom shower. But we should worry about truckloads and containers full of scenery and costumes that you can smell before you can see, materials with visible patches of mold growing on them, or shops located in water-damaged buildings.

What Are They?

Molds are living organisms that can grow whenever temperatures are above freezing and humidity is a little high. The higher the humidity and temperature, the faster they grow. Almost everything is food for some types of mold: fabric, wood, plasterboard, paper, paint, plastics, and more. Molds can even grow on the dust and lining material in ventilation ducts. Once in the ventilation system, molds can invade whole buildings.

How Do They Hurt Us?

Molds release vast numbers of spores, which can get into the air when people try to clean them up or disturb them in some way. Once excessive amounts are airborne, molds can cause serious health effects. These include the following:

- *Irritation* of the respiratory tract and eyes can be experienced by all workers if exposure is high enough.
- *Allergy* symptoms only occur in people who have developed a reaction to particular molds. The level of exposure at which symptoms occur varies greatly among individuals. Overexposure can cause people who were not previously sensitive to molds to develop the allergy. Once they exist, allergies usually are permanent and may worsen with each additional exposure. Symptoms range from eye itching and hay fever to life-threatening asthma and anaphylactic shock.
- *Toxicity.* Some molds, just like the mushrooms to which they are related, produce powerful poisons. One toxic mold, called *Stachybotres chartarum,* was once used in germ warfare. It can cause serious neurological damage and death. It is suspected as the cause of sudden deaths in infants living in moldy environments.
- *Infection.* Some molds can take up residence and grow in the lungs or other bodily organs. This usually occurs only in people with compromised immune systems, such as those on chemotherapy or who have AIDS.
- *Immune system damage.* The drug cyclosporin, used to depress the immune systems of organ transplant patients, is a mycotoxin. A number of species of molds produce this and other toxins that are capable of damaging immune function.

CLEANING UP MOLD

An occasional moldy costume can be carefully put into a washing machine with bleach. A small patch of mold on a wall can be cleaned up with a 1:10 bleach solution. But it is not safe to disturb mold in quantity. Boxes full of moldy materials and infestations on building walls or on scenery cannot be addressed by applying bleach or other disinfectants. Merely killing mold is not the answer. Dead mold particles can still cause irritation, allergies, and toxicity! Instead, take the following steps:

1. Do not attempt to clean up heavy mold infestations yourself. Disturbing mold can result in serious exposures to spores and dust.
2. Alert fellow workers. All workers must be warned to avoid exposure. Those with allergies or depressed immune systems must be especially careful.
3. Ask your employer to have the mold identified, to be certain that none of the highly toxic molds are present. Many hospitals have mycologists on staff who can identify molds. There are a number of commercial labs that will also do this. One is:

Dr. Chin Yang
P&K Micro.—Unit L
1950 Old Cuthbert Road
Cherry Hill, New Jersey 08034
(609) 427-4044
PKMICROBE@aol.com

4. Get professional advice from an industrial hygienist or your health and safety officer about how to clean up. Some jobs require a professionally trained abatement crew.
5. Mold needs to be controlled. Once molds are cleaned up, procedures must be used to insure the problem does not recur. Controls include:
   - Cleaning fabrics and surfaces with a HEPA vacuum
   - Eliminating carpeting, drapes, or other unnecessary materials that act as repositories for mold and spores
   - Providing environmental controls. Molds become dormant when the humidity and/or temperatures are low. An engineering survey can be done to determine how to repair and upgrade the ventilation system to control temperature and humidity.

OTHER BIOLOGICAL HAZARDS

Hantavirus

This virus is now known to exist throughout the entire North and South American continents. And more than one type of rodent can be infected. The disease can be contracted when dust, containing small virus particles from rodent urine, are inhaled during cleaning or disturbing of nesting materials. The disease is between 40 and 60 percent fatal.

If you see large areas of rodent nests and droppings, do not clean them yourself. There are commercial services that are trained to clean up biological wastes.

It is safe to remove an occasional small dead animal, a mouse, or a bird nest. Make a 1:10 bleach-and-water solution while wearing splash goggles and gloves. Spray the solution on the refuse until it is soaked, and place in double plastic bags without raising dust.

Pigeon Waste

Large amounts of pigeon waste can collect in a very short time if pigeons have access to buildings. This is a common problem in temporary location shops and summer theater locations. But wherever it is, pigeon waste and nesting materials are known to be sources of a number of disease organisms. Included are:

- Allergenic hypersensitivity pneumonitis, or "pigeon breeders disease." Frequently misdiagnosed as pneumonia, some forms of this disease can lead to permanent impairment. Once sensitized, a person may react to extremely low or even unmeasurable concentrations of dust containing pigeon material.
- *Cryptococcus neoformans* (a yeast) may cause mild pulmonary infections or skin lesions in healthy people or encephalitis in immune-deficient individuals. This disease affects 6 percent of all HIV-infected persons. Untreated, the fatality rate is high.
- *Histoplasma capsulatum* (a fungus) grows in soil near droppings. Symptoms occur three to eighteen days after exposure. Its severity can range from asymptomatic to a flu-like respiratory illness. A small percentage of people develop a form of the disease that resembles tuberculosis that can worsen over months and years. The rarest form of the disease is histoplasmosis, a potentially fatal infection.
- *Aspergillus fumigatus* (a fungus) is especially dangerous to immune-deficient people. Healthy people exposed to it can develop allergic aspergillosis, a severe hypersensitivity lung condition. Pigeon waste can support growth of this and many other hazardous molds (see Molds above).
- *Chlamydia psittaci* (a bacterium) causes a disease called psittacosis. Its severity can range from asymptomatic to a severe and fatal pneumonia.

Legionnaires' Disease

Any standing water can begin to grow the organism that causes Legionnaires' disease. The organism has been found in wet grinding equipment and has been transmitted by mist rising from a Jacuzzi. Any standing water in shops should be cleaned up before trouble strikes.

Tetanus, Salmonella, etc.

There are many diseases transmitted by animals and unclean conditions. None of them are a hazard if the shop is kept clean, dusts are not raised during cleaning, and 1:10 bleach is used when a disinfectant is needed.

16 Carpentry

Pine, plywood, and composition boards are the primary set construction materials. However, virtually any type of wood may end up in prop- and scene-making if old hardwood objects or furniture are modified, if exotic woods are used, and the like. Many people consider dust from wood as nothing more than a nuisance. It is far more than that.

WOOD DUST HAZARDS

Wood dust has caused countless fires and explosions. A spark or static discharge is sufficient to detonate fine, airborne sawdust. In addition, some wood dusts cause allergies, some are toxic, and others contain highly toxic pesticides, preservatives, flame retardants, and other treatments. Some trees deposit significant amounts of toxic silica in their heartwood. It has also been established that certain types of cancer are related to wood-dust exposure.

Why Is Wood Toxic?

Trees are just big plants. Like all plants, they are chemical factories that produce hundreds of substances, some of which are capable of fighting insects, resisting fungus, and keeping other trees from invading their space. Most of these wood components have never been tested for their biological effects on humans. Examples of toxic wood components include the poisonous alkaloids in yew and oleander and strychnine found in some types of boxwood.

In addition to toxic chemicals, many of the chemicals in wood can cause irritation and allergies. These chemicals are clearly also the cause of a variety of occupational illnesses that are seen in wood dust–exposed individuals.

Diseases from Wood Dust

Irritant and allergic dermatitis can be caused by exposure to wood dust. Respiratory-system effects, such as damage to the mucous membranes and dryness and soreness of the throat, larynx, and trachea, can be caused by some woods. Serious lung diseases, such as asthma and alveolitis, can affect workers exposed to sawdust.

Epidemiologic studies of furniture workers exposed to hardwood dust have indicated an excess of lung, tongue, pharynx, and nasal cancer. Hardwood dusts are definitely associated with cancer, and limited results from studies on workers exposed to soft woods indicate that both types of wood are implicated.

Cancers of the nasal cavity and sinuses are the most prevalent of the types of cancer associated with wood dust. A twelve-country survey showed that an astonishing 61 percent of all such cancer cases occurred among woodworkers.[1] It is estimated that about 2.5 percent of woodworkers will develop this kind of cancer, usually after a latency period of up to forty years.[2,3]

Exposure Standards

Carpenters can protect themselves from wood dust hazards by keeping their exposure low. The American Conference of Governmental Industrial Hygienists (ACGIH) has set a Threshold Limit Value (eight-hour time-weighted average) for wood dust that considers these effects. The current standard is 5 milligrams per cubic meter (mg/m^3).[4]

In 1989, OSHA also set Permissible Exposure Limits (PELs) for wood dust of 5 mg/m^3 and 2.5 mg/m^3 for western red cedar. These limits and hundreds of others were vacated by the courts. However, the major wood industries recognize that 5 mg/m^3 is a reasonable limit and have entered into agreements with OSHA to adhere to this limit. The problem is that our industry has not agreed to this limit, and so there currently is no OSHA PEL for wood dust in our shops. The only way OSHA can cite for wood dust exposure in our shops is to invoke the General Duty Clause.

MSDSs Needed

In 1994, OSHA published in the Federal Register (59 FR 17478-9) some technical amendments to the Hazard Communication Standard. The amendments clearly state that the law applies to wood products that are to be processed in a

---

[1]J. H. Wills, "Nasal Carcinoma in Woodworkers: A Review." *Journal of Occupational Medicine* 24(7), 527, 1982.

[2]G. Swanson and S. Belle, "Cancer Morbidity Among Woodworkers." *Journal of Occupational Medicine* 24(4), 315, 1982.

[3]See also the 1991 ACGIH Documentation of TLVs (Bibliography).

[4]Actually, the limits at the time of writing this book are 1 mg/m^3 for hardwood and 5 mg/m^3 for softwood. However, ACGIH has published a notice that it intends to change the hardwood standard to $5m/m^3$, and I anticipate this will happen.

manner which creates wood dust. Previously, the wood industry did not accept this interpretation of the OSHA rule and did not supply MSDSs for all wood and wood products. Now MSDSs should be readily available on all wood products, and the MSDSs must be on file at the job site (29 CFR 1910.1200(b)(6)(iv).

WOOD TREATMENTS

If the wood is chemically treated, the hazards of these chemicals also may be addressed on MSDSs. Almost every imported wood and most domestic wood in the United States and Canada has been treated with some kind of additive. These can include fire retardants, pesticides, and preservatives. These chemicals range in toxicity from relatively safe to highly toxic. The most hazardous of the additives are pesticides applied to wood intended for use in contact with the outdoor elements.

Three common outdoor-use wood preservatives are pentachlorophenol (PCP) and its salts, arsenic-containing compounds, and creosote. These three types of preservatives are associated with cancer, birth defects, and many other hazards. The only time such woods should ever be used is for permanent outdoor installations. Even then, the wood should be delivered precut. OSHA regulations would require personal monitoring of workers cutting wood preserved with arsenic compounds or PCP.

PLYWOOD AND COMPOSITION BOARDS

Wood glues and adhesives used in plywood, pressboards, and many other wood products may contain urea-formaldehyde or phenol-formaldehyde resins. These glues outgas formaldehyde, which is a strong eye and respiratory irritant, an allergen, and a suspected carcinogen.

Some manufacturers are turning to other adhesives, such as urethane adhesives. It is hoped these binders will be less hazardous, but hot saw blades can release small amounts of toxic substances from these resins as well (see chapter 13, Plastics and Adhesives).

WOOD SHOP PRECAUTIONS

Ventilation

In order to meet the current wood-dust standards, it is necessary to equip woodworking machines with local exhaust. The best method is to connect each machine to a central collection system that expels the dusty air from the shop rather than filtering the air and returning it. For shops unable to afford such systems, portable collectors and/or bag systems that can be moved from machine to machine may suffice. But fine dust will escape these systems, and extra cleanup of workplace surfaces and some respiratory protection will be needed.

Respiratory Protection

Respiratory protection may be used to reduce employee exposure to wood dust (see chapter 9, Respirators). Most wood dust is in a particle-size range that is easy to capture with the 95-series respirators. The exception would be fine sanding and the fine dusts from some kinds of composition boards, such as maximum density fiberboard (MDF).

Machine Safety

Woodworking-machinery rules require guards at the point of operation, cutoff switches, lockouts, kickback guards, returns, and other precautions (1910.213). For example, hand-fed ripsaws require three components to a guarding system: the barrier, the spreader, and the anti-kickback fingers or dogs. The drive belts and other moving parts that transmit mechanical power to the machines must also be enclosed and guarded (1910.219).

There are general rules for safe use of hand and portable power tools (1910.242(a)) which require that employers shall be responsible for the safe condition of tools and equipment used by employees, including tools and equipment that may be furnished by employees. Guarding and safety features of each kind of portable power tool is also specified (1910.243).

Sometimes, compressed air is used to clean surfaces or clothing. This is a bad idea with respect to dust control. OSHA allows it only when the air pressure is reduced to less than thirty pounds per square inch and with protective equipment, such as eyewear with sideshields (1910.242[b]).

Many shops have abrasive grind wheels. These machines must have their guards in place, their tool rests must be adjusted to one-eighth inch from the wheel, and breakout plates (adjustable tongues) should be a maximum of one-quarter inch from the wheel. It is common to see, instead, grind wheels stripped of their guards and misused—including evidence that the side of the wheel has been used for grinding. This weakens the wheel and increases the risk that it will shatter. Violation of OSHA rules regarding grind wheels are considered serious, because people have lost fingers caught between tool rests and wheels, and others have been blinded and even killed by shattering abrasive wheels.

OTHER WOOD SHOP HAZARDS

Vibrating Tools

A significant number of persons who use vibrating tools now are known to be at risk from a more permanent condition commonly called "white hand," "dead fingers," or "vibration syndrome." This disease can lead to severe pain, ulcerations, and even gangrene in some cases.

Noise

Saws, planers, routers, sanders, and the like can easily produce a cacophony of ear-damaging sound waves. Poorly designed ventilation systems can also produce hazardous amounts of noise (see chapter 7).

GENERAL PRECAUTIONS FOR WOODWORKING

- Prevent fires by providing good shop ventilation, dust collection and control, sprinkler systems or fire extinguishers, emergency procedures and drills, proper storage of any flammable products, and by banning smoking.
- Only use machinery that meets all of OSHA's standards for guarding and safety features. Train workers on each type of tool, and keep training records. The manufacturers of table saws and other woodworking machinery often have training tapes and materials available.
- Equip all woodworking machinery with local exhaust dust collection systems. Ideally, these systems should vent to the outside, rather than return air to the shop.
- Obtain MSDSs on all wood products, glues, and other materials. Use MSDSs to choose the safest products.
- Avoid wood treated with PCP, arsenic, or creosote. Some of arsenic-preserved wood can be identified by its greenish color.
- Prevent hearing damage by purchasing quiet machines using dampening equipment, such as mufflers or rubber mounts. Keep machines well-oiled and maintained. Use earplugs or muffs if needed. Have a baseline hearing test and periodic audiograms as often as your doctor suggests or your workplace hearing-protection plan requires.
- Prevent vibration syndrome by using tools that are ergonomically designed and produce low-amplitude vibrations, working in normal and stable temperatures (especially avoiding cold temperatures), taking ten-minute work breaks after every hour of continuous exposure, and not grasping tools too hard.
- Wear eye protection rated for impact, with sideshields, at all times in the shop. Even if you are not using a power tool, someone near you may be.
- Wear a dust mask or cartridge respirator with a particulate filter when dust cannot be controlled by ventilation. Most particulate filters will be sufficient, unless very fine dusts are created. In these cases, use the 100 series of filters (see chapter 9, Respirators).

- Wear protective clothing to keep dust off your skin. Wear gloves when handling woods known to be strong sensitizers.
- Practice good hygiene. Wash and shower often. Keep the shop clean and free from sawdust. Vacuum rather than sweep dusts.
- Follow all rules for use of solvents when using solvent-containing paints, glues, and similar products (see chapter 11, Solvents).
- Be prepared for accidents. Know your blood type, and keep up your tetanus shots. Keep first-aid kits stocked. Post and practice emergency procedures. Follow all medical surveillance rules below.

MEDICAL SURVEILLANCE FOR WOODWORKERS

- For occupational health problems, consult a doctor who is board certified in occupational medicine or one who is familiar with wood-related illnesses.
- Have baseline lung-function tests done early in your woodworking career. Then have your physician compare this test with subsequent pulmonary function tests done in your regular physical examinations in order to detect lung problems early.
- Have your physician pay special attention to your sinuses and upper respiratory tract. Report symptoms like nasal dripping, stuffiness, or nosebleeds.
- Have a baseline audiogram done early in your career as a carpenter. Have your physician compare this test with subsequent audiograms to see if your hearing is being affected by your work.
- Be able to give your doctor a good occupational history. Always be prepared to provide your doctor with information about the chemicals you use and your work practices. Keep records of recurring symptoms and chemical/wood exposures.
- Watch for symptoms of vibration disease, such as fingers getting white or abnormal pain in the hands during cold weather. Report these symptoms to your doctor.
- Suspect that a health problem may be related to your work if the condition improves on weekends or during vacations.

17 Welding

Many methods of welding are used in theater. Those most often used are oxyacetylene welding, simple arc welding, metal inert gas or MIG welding, tungsten inert gas or TIG welding, and plasma cutting. All these methods involve high heat from either burning gas or an electric arc.

All types of welding can be extremely hazardous. Nowhere in theater shops is there a greater potential for fires, burns, electrical hazards, explosions, and health hazards than from welding. And yet, welders in many shops, including those in university theater programs, often have no welding credentials. It is scandalous that schools that require their drama teachers to hold college degrees will hire welding instructors to teach this dangerous craft who have never had a single day of formal training.

BASIC RULES AND REGULATIONS

Safe welding requires knowledge, training, and compliance with health and safety codes and regulations.

The Standards

Welders and those responsible for administering or setting up theater welding shops should be thoroughly familiar with the American National Standards Institute's and American Welding Society's joint standard, ANSI/AWS Z49.1-94, "Safety in Welding and Cutting and Allied Processes," the National Fire Protection Association (NFPA) codes, the OSHA General Industry Standards from 1910.251-257 and the Construction Standards from 1926.350-351 (see Bibliography).

Training

Welders should be trained and certified to do the types of welding done in the shop. For example, the American Welding Society (AWS) considers 125 to 150 hours of professional training necessary to qualify for oxyacetylene, oxyfuel gas welding, brazing, and flame cutting (which are taught as a unit).

Welders who teach students or supervise other welders should have additional training in health and safety. Seminars in welding health and safety also are provided by the AWS (see address in Appendix).

Training and Liability

Training is also needed to protect the employer's liability. For example, if a set's welded elements break apart and cause an accident, the welder's credentials will be a key element in any resulting legal action.

Welding certification programs are designed to teach welders to make good welds—welds that will have the proper characteristics and strength. Welds that look good can have hidden flaws. Scenery should never have load-bearing elements that were joined by uncertified welders. If this is not possible, hire a certified welding inspector to use special equipment to check each load-bearing weld before sets are used.

General Fire Safety

Over 4 percent of all industrial fires are started by welding sparks. These sparks are actually molten globules of metal that can travel up to forty feet and still be hot enough to ignite combustible materials. Fire extinguishers must be on hand in welding shops. If sprinkler systems also are installed, check to be sure the type of system and the location of sprinkler heads is compatible with welding activities. (Imagine the results if an electric arc welder were suddenly deluged with water.)

Safe Shop Design

Shops should be prepared to comply with the OSHA standards for welding fumes and gases. Shops must be carefully constructed to eliminate any place where sparks may lodge and smolder. Cracks in walls or crevices between floorboards are examples of dangerous conditions.

Other design elements should include:

- Floors and furniture made of noncombustible materials or coated with fire-resistant paints
- Walls and other surfaces coated with dull finishes such as can be obtained from special nonreflective paints

- Ventilation systems capable of removing welding fumes
- Eyewash stations, first-aid equipment, proper storage units for welding protective gear
- Portable welding screens or curtains to isolate welding areas
- Fire-protection systems appropriate to the space
- Proper storage and anchorage for compressed gas cylinders

A commonly violated shop-design standard is OSHA's requirement that combustibles should not be within thirty-five feet of welding operations. Yet, I have seen plans submitted by prominent architects to major schools that place welding activities in wood shops where sawdust and combustibles will be present. In another case, a university theater welding shop was near an open elevator shaft. Welding sparks actually started fires in oily residue at the shaft's bottom on five or six separate occasions before the university relocated this operation.

Welding shops should be for welding only. Many other activities are incompatible with welding. One especially dangerous activity is spray painting. Solvent-containing overspray mists can explode if ignited by a welding spark. Even water-based paint mists may contain enough combustible solids to ignite under certain conditions.

On-Site Welding

The same thirty-five-foot OSHA rule for combustibles applies to welding done on stages or on location. If combustibles cannot be removed, OSHA requires the employer to have a fire watcher who has a fire extinguisher at the ready all during the welding and for a full half hour after the last torch is off. Even then, highly combustible substances within thirty-five feet are prohibited. Examples OSHA gives for combustibles that must be kept thirty-five feet away from on-site welding operations include "paper clippings, wood shavings, or textile fibers" (1910.252[a][v]).

Welding operations on stage are particularly tricky. All other activities on stage should cease during welding. All curtains and drops should be drawn away from the welding area, combustibles should be removed or covered with fire-resistant tarps, and a fire watcher should be present and remain for a half hour after welding ceases.

Good Housekeeping

One of the simplest and most effective safety practices is keeping welding shops scrupulously organized and clean. Nothing that is not necessary to the welding shop should be stored in it. Organize the space so that nothing is on the floor that welders can trip over. Welders' vision is often limited by faceshields or goggles.

Electrical Safety

Most shocks caused by welding equipment are not severe. But under the right conditions, they can cause injury or even death. Mild shocks can cause involuntary muscle contractions leading to accidents, and moderate amounts of current directed across the chest may stop the heart. Here are some basic ways to avoid these hazards:

- Use only welding equipment meeting national standards, such as those of the National Electrical Manufacturer's Association (NEMA)
- Follow exactly all equipment-operating instructions
- Keep clothes dry (even from excessive perspiration), and do not work in wet conditions.
- Maintain all electrical connections, cables, electrode holders, etc., and inspect each before starting to weld.

Compressed Gas Safety

Compressed gas cylinders are potential rockets or bombs. If mishandled, cylinders, valves, or regulators can break or rupture, causing damage as far as one hundred yards away. The OSHA rules for compressed gases (1910.101) reference the standards of the Compressed Gas Association. There are three basic types of hazardous gases:

- *Oxygen.* It will not burn by itself, but ordinary combustible materials like wood, cloth, or plastics will burn violently or even explode when ignited in the presence of oxygen.
- *Fuel gases.* Acetylene, propane, and butane are some fuel gases. They are flammable and can burn and explode.
- *Shielding gases.* These are used to shield processes such as MIG and TIG welding and include argon, carbon dioxide, helium, and nitrogen. They are inert, colorless, and tasteless. If they build up in confined spaces such as enclosed welding areas, they replace air and can asphyxiate those in the area.

There are many regulations and standards established to protect welders from the hazards of compressed gas cylinders. Some basic rules include the following:

- Accept only cylinders approved for use in interstate commerce. Do not remove or change any numbers or marks stamped on cylinders.

- Cylinders too large to carry easily may be rolled on their bottom edges, but never dragged.
- Protect cylinders from cuts, abrasions, drops, or striking each other. Never use cylinders for rollers, supports, or any purpose other than intended by the manufacturer.
- Do not tamper with safety devices or valves.
- Return empty cylinders to the vendor. Mark them "EMPTY" or "MT" with chalk. Close the valves and replace valve protection caps.
- Always treat cylinders as though full (even when empty), and handle them with care. Accidents have resulted when containers under partial pressure have been mishandled.
- Secure cylinders by chaining, tying, or binding them, and always use them in an upright position.
- Store cylinders in cool, well-ventilated areas or outdoors in vertical positions (unless the manufacturer suggests otherwise). The temperature of a cylinder should never exceed 130°F. Store oxygen cylinders separately from fuel cylinders or combustible materials.

HEALTH HAZARDS

Hazards to welders' health are less obvious than welding safety hazards, and they vary among different types of welding. In general, the hazards are radiation, heat, noise, fumes and gases from welding processes, and gases from compressed cylinders.

Radiation

Radiation generated by welding takes three forms: visible, infrared, and ultraviolet.

Visible light is the least hazardous and most noticeable radiation emitted by welding. Intense light usually only produces temporary visual impairment.

Infrared radiation (IR) is produced when substances are heated until they glow—as during welding, cutting, brazing, or soldering. IR can cause temporary eye irritation and discomfort. Repeated exposures can cause permanent damage to the retina and perhaps to the lens of the eye (cataract). Chronic IR eye damage of these types occurs slowly, without warning.

Ultraviolet (UV) is the most dangerous of the three types of radiation. Eye damage from UV, often called a flash burn, can be caused by even a few seconds' exposure. Symptoms usually do not appear until several hours after exposure. Severe burns become excruciatingly painful, and permanent damage may result.

UV can also damage exposed skin over time. Chronic exposure can result in dry, brown, wrinkled skin and may progress to a hardening of the skin called keratosis. Further exposure is associated with benign and malignant skin tumors.

Radiation Eye Protection

The type of eyewear needed to protect welders from these hazards is found in figures 3 and 4 in chapter 7. Workers should not be doing other types of work near welding operations. If they must, they also must be wearing eyewear to protect against glare and UV at a distance. Often called "visitors' eyewear," every shop should have a few pairs available if someone must be in the shop during welding operations.

Welders' eyewear should also be rated for impact. A common injury occurs when welders raise their hoods to inspect a weld and slag pops off unexpectedly.

Heat

Heat can harm welders by causing burns (from IR radiation to the skin or from hot metal) and by raising body temperature to hazardous levels, causing heat stress.

Noise

Noise can damage a welder's hearing. Fortunately, most welding processes used in theater produce noise at levels well below the OSHA Permissible Exposure Limits. (Air carbon arc cutting and hammering on metal are exceptions.) Welders should only wear fire-resistant earplugs. Several cases of eardrum damage have been reported when an overhead spark fell into an ear canal that was either unprotected or contained a combustible plug.

Fumes

Fumes and gases are produced during the welding process. They sometimes can be seen as a smoky plume rising from the weld. Fumes are created whenever metal is melted, just as water gets into the air when it evaporates. Once this metal vapor is released, it reacts with air to form tiny metal oxide particles called fumes.

Many occupational illnesses are associated with substances found in welding fumes and gases, including metal fume fever (with symptoms similar to flu and usually from zinc fumes) and a variety of chronic lung diseases, including chronic bronchitis. Lung and respiratory-system cancer are associated with metal fumes such as chrome, nickel, beryllium, and cadmium. These metals are usually encountered when junk metals are welded. Junk-metal welding is always a dangerous proposition, since the composition of the alloy is unknown. In addition, welding painted metals may expose welders to lead fumes from paint. Metal priming paints are still exempt from the lead paint laws.

GENERAL PRECAUTIONS

Dressing for the Part

Wear clothing appropriate to the job. Pants and long-sleeved shirts made of wool or heavy cotton fabrics insulate welders from temperature changes and resist burning. Wool also emits a strong warning odor when heated or burned. Any light cotton fabric must be treated with a flame retardant. Never wear synthetic fabrics, which melt and adhere to the skin when they burn. Pants and shirts should not have pockets, cuffs, or folds into which sparks may fall. Holes in fabrics or grease stains on clothing are not acceptable.

A quality leather boot with a minimum six-inch-high top should be worn. A skullcap or cap with a bill should be worn over hair. A range of gloves should be available, from light-welding gloves for jobs that need dexterity to full-leather-gauntlet gloves for arc welding and plasma cutting. Leather aprons, leggings, spats, and arm shields may be needed for some welding jobs.

Safety glasses with sideshields should be worn at all times in the shop. They are required under welding helmets or grinding shields. The safety glasses may be prescription. Contact lenses are acceptable, but the drying effects of heat and radiation may make them more difficult to use.

Shields and goggles for protection against radiation for various types of welding are shown in figures 7.3 and 7.4 in chapter 7. These must be kept in good condition. A scratched lens will permit radiation to penetrate, and it should be replaced.

A selection of hard hats and welding helmets should be available. The required OSHA training should be designed to prepare welders to choose the right equipment for each job.

Ventilation

Welding shops should have local exhaust ventilation systems, such as downdraft tables or flexible duct fume extractors, to capture welding emissions at their source. These local exhaust systems should be combined with dilution systems to remove gases and fumes that escape collection and to reduce heat buildup. A simple exhaust fan and makeup air supply may suffice for open-area welding of low-toxicity metals such as mild steel. Enclosed MIG and TIG welding booths will need floor-level dilution systems to prevent layering of inert gases.

Respiratory Protection

Ventilation, not respiratory protection, is the first choice for theater welding jobs. No single air-purifying respirator will protect welders against all the contaminants in welding plumes. HEPA filters will capture most metal fumes, but they offer no protection from gaseous contaminants. Some air-supplied respirators can provide welders with fresh air, but these are complex pieces of equipment that are effective only when maintained properly and only when their users receive proper training. (See chapters 9 and 10 on respiratory protection.)

18 From Start to Strike

Construction of set components, installation of scenery, rigging, setting lights, and striking are traditional activities in theater and entertainment work. And now, the traditional ways in which people have done this work must change, because OSHA had changed the regulations that apply to this work. Full harnesses must replace belts for fall protection. Formal classroom training must be provided and documented for workers on jobs that involve fall protection or the use of scaffolds or powered lifts.

It now is an OSHA violation for that untrained, eager amateur to climb the rigging, step onto a scaffold, or drive a powered lift. All theaters, companies, or theater schools must either have trained and/or certificate-holding qualified people on staff or get busy and send their employees to the safety schools that provide this training.

FALL PROTECTION

OSHA has two separate rules for fall protection that apply to theatrical work: the construction industry standard and the general industry standard.

1. *The construction industry standard* for fall protection (1926.500-503) applies during set building and erecting, setting of lights, and rigging. This rule requires that workers must be tied off and in a body harness when there are any conditions that make it possible for them to fall more than six feet. Appendix E of this OSHA standard provides a prototype program, which can be adapted by employers to meet the requirement for a written program.
2. *The general industry standard* for fall protection (1910.23) applies when construction is finished and performers and other regular employees are on site. Under the general industry standard, fall protection is needed at four feet instead of six feet.

From Harness to Anchorage

In the past, belts and sit/climbing harnesses were used for this kind of work. They now cannot be legally used for this purpose. Accidents showed that these kinds of belts can cause serious back injuries and even death when they break a person's fall.

Today, a harness must be attached by a sliding D ring to a lanyard that has a shock absorber and is tied off at an anchorage point. All components and connections, from anchorage to harness, must be capable of withstanding five thousand pounds of tensile force (more than sufficient to stop a heavy person falling at the end of a lanyard).

Experienced theater riggers will immediately see a problem with this rule. They know that many theaters have loading bridges with rickety railings on the onstage side and nothing but a thin chain on the offstage side. They have seen catwalks over the seating area where nothing within reach looks solid enough for anchorage. Theaters must hire engineers and rigging consultants to correct these problems and provide proper anchorage.

Rails and Nets

Another legal way to address fall hazards is to use guardrails or safety nets. The precise height, location, and strength of guardrails and details about safety nets are described in 1926.501(b) and (c) or in 1910.23. The rails and nets must be constructed precisely as OSHA requires, since they are engineered to enable them to withstand the force of a heavy person failing. Here is the general industry description of wooden rails:

> (1910.23(e)(3)(i) For wood railings, the posts shall be of at least 2-inch by 4-inch stock spaced not to exceed 6 feet; the top and intermediate rails shall be of at least 2-inch by 4-inch stock. If top rail is made of two right-angle pieces of 1-inch by 4-inch stock, posts may be spaced on 8-foot centers, with 2-inch by 4-inch intermediate rail.

Toe boards are required under the railings, and there are OSHA specifications for metal railings as well. For certain types of elevated platforms, OSHA allows special restraint systems that will keep workers away from the edges of the platform (see 1926.502(f) to (h)).

The Stage: A Special Problem

In a January 28, 1997 letter of interpretation entitled, "Fall protection for the entertainment industry under the OSHA Act of 1970," OSHA made it clear that the lip of the stage does not need to be guarded, but that the fall protection regulations apply, and some other solution will be needed. They said:

> OSHA is concerned with the safety and health of all workers in the entertainment industry. Although OSHA recognizes it is not appropriate to put guardrails at the edge of stages, theatrical employees need to be protected from all occupational safety and health hazards. The fall protection standards for general industry (found in Subpart D of 29 CFR at 1910.21 through 1910.32) as well as the personal protective equipment standards (found in Subpart I of 29 CFR at 1910.132 through 138) are the appropriate standards for your situation.[1]

It is clear from this statement that OSHA considers the stage a general industry site under 29 CFR 1910, but that the edge of the stage does not need a guardrail. However, there are ways to reduce this risk that do not conflict with artistic needs. For example, only allowing professionals on the stage, assigning crossover paths, and planning blocking that keeps people from areas of danger is helpful.

For another example, 1910.23(e)(8) for skylight screens might be adapted to put netting over the orchestra pit.

> .23(c)(8) Skylight screens shall be of such construction and mounting that they are capable of withstanding a load of at least 200 pounds applied perpendicularly at any one area on the screen. They shall also be of such construction and mounting that under ordinary loads or impacts, they will not deflect downward sufficiently to break the glass below them [or hit the orchestra players?]. . . .

The netting might be an especially useful method for stages in schools, where amateurs may be at risk or where stunts or acrobatics may put orchestra members as well as performers at risk.

Stairs

Stairs also must be guarded under both the general industry and the construction standards. The rule, found in 1910.24(d)(1), says:

> Every flight of stairs having four or more risers shall be equipped with standard stair railings or standard handrails as specified in paragraphs (d)(1)(i) through (v) of this section. . . .

[1]http://www.osha-slc.gov/OshDoc/Interp_data/I19970128.html

A complete description of all these specifications will not be included here, but set designers and builders must be familiar with them and arrange to have them in place even when stair units are used during set construction and after.

Training

Any person expected to wear fall protective equipment, to be protected by guardrails, or to work at heights or near leading edges must have training. Training must be provided by persons who themselves are competent to train. The subject matter required for training is specified by OSHA, and the employer must prepare a signed and dated, written certification record of the training.

SCAFFOLD REGULATIONS

Both the general industry and construction standards regulate the use of scaffolds (1910.28 and 1926.451). These laws have changed recently. One of those changes involves the definitions of two types of personnel, defined by OSHA as "competent" and "qualified."

- *Qualified person* means one who, by possession of a recognized degree, certificate, or professional standing, or who by extensive knowledge, training, and experience has successfully demonstrated his or her ability to solve or resolve problems related to the subject matter, the work, or the project.
- *Competent person* means one who is capable of identifying existing and predictable hazards in the surroundings or working conditions that are unsanitary, hazardous, or dangerous to employees, and who has authorization to take prompt corrective measures to eliminate them.

Qualified people can design scaffolding for a job, choose equipment, and make similar decisions for a specific project. Each person who will work on the scaffold must be trained by someone qualified in the subject matter to recognize the hazards associated with the type of scaffold being used and to understand the procedures to control or minimize those hazards. OSHA specifies the subject matter that must be covered in training.

It is the competent person who is the workhorse on site. The competent person is responsible for seeing that all users understand the complex rules about planking of scaffolds, how to ascend and descend properly, when mobile scaffolds can be moved, use of rails and fall protection, use of hard hats, control of tools and objects that may fall on workers below, electrical safety (metal scaffolds can be energized by defective power equipment), etc.

The competent person must also train each employee who is involved in erecting, disassembling, moving, operating, repairing, maintaining, or inspecting a scaffold. OSHA specifies the subject matter in which these employees must be trained.

The most important part of the competent person's position is that he or she must have "authorization to take prompt corrective measures to eliminate them." The employer must keep the training records and determine when retraining is necessary.

POWERED INDUSTRIAL TRUCKS AND LIFTS

This rule (1926.178(1)) requires operators of forklifts, platform lift trucks, and other powered lifts, including Genie® truck lifts, to be trained before they operate them. Training must consist of both classroom and practical training in proper vehicle operation, the associated hazards, and requirements of the OSHA standard.

The lift errors most often seen are using Genies or other personnel lifts with the outriggers removed, moving the lifts while the basket is raised and a person is in the basket, and using them on uneven ground without first leveling the lift.

RIGGING SYSTEMS

Operating stage rigging systems and attaching scenery to these systems are the most hazardous parts of theater production. Stage rigging systems include hemp-rope sandbag-balanced systems, counterweight systems using hemp and wire rope, dead hung or nonmovable rigging, and remote-controlled electrical winch systems.

Discussing detailed hazards and precautions for each of these systems and the methods for attaching scenery to them is beyond the scope of this book. Fortunately, a considerable body of literature already exists on this subject (see Bibliography), and the construction and maritime rigging industries use techniques applicable to technical theater.

Personnel responsible for rigging should take refresher courses and seminars periodically, since regulations and equipment change frequently. Two good sources are:

Sapsis Rigging
233 North Lansdowne Avenue
Lansdowne, Pennsylvania 19050
(800) 292-3851
Bsapsis@aol.com

Rigging Seminars
P.O. Box 486
Bedford, Indiana 47421
(812) 278-3123
rigging@riggingseminars.com

Some general recommendations, however, can be made to improve rigging safety procedures.

- Arrange work schedules so that you undertake rigging and flying when no other activities are taking place on stage.
- Permit only authorized and trained personnel to rig scenery and operate rigging systems.
- Hold a safety-and-strategy meeting at the beginning of each work period for the entire crew. If students, trainees, or new hires are present, review safety procedures and warning-call terminology.
- Define each individual's job. Everyone should know precisely what his or her responsibilities are as defined by a job description, the *Handbook of Theatrical Apprentices* (see Bibliography), or some other suitable set of formal guidelines.
- Establish lines of command. For example, the crew head or technical director should be the only one who calls instructions to the grid crew.
- Order and insist upon periods of complete silence on stage during especially hazardous operations, such as when an arbor is being loaded or unloaded. Never allow noise to reach levels sufficient to cover warning calls.
- Before going onto the grid, crew members should make sure that nothing on their person can fall to the stage. Empty pockets, remove or secure glasses, jewelry, hair ornaments, etc. Secure any tools to be used on the grid to the worker with a safety line.
- Practice appropriate fall protection while working on the grid, bridges, lifts and the like (see above).
- Wear protective clothing appropriate to the type of work. Wear hard hats whenever overhead rigging is in process, and wear rubber-soled shoes or boots at all times.
- Never drop any scene component or object from the grid to the floor. Raise or lower objects on lines. Ropes and electrical cord should be coiled before carrying or lowering them to stage level.
- After rigging is finished, check it and have it approved by a qualified person. Since faulty rigging can cause such serious accidents, paying a private inspector's fee (usually between $500 and $1,000) is worthwhile to avoid accidents and protect liability.
- Report any defective or worn equipment immediately, and seek replacements.

- Do not use any equipment or materials whose load-bearing capacity is not known.
- Report any accidents, incidents (near misses), or foreseeable problems immediately.
- Once rigging is in place and being used, report and discuss any problems during rehearsal notes.
- Strike show rigging—including extraneous hardware, batten extensions, and other attachments—at the end of each production's run.

IRREGULAR SURFACES

Accidents and falls commonly occur when walking surfaces are uneven, raked, or contain stairs, pits, traps, or other changes in elevation. For example, there is now evidence that dancers' accident rates rise when they work on raked stages.[2]

Obviously, guardrails can't be installed around changes in elevations during performances. For this reason, stunt falls through pits and traps must only be done by people trained and rehearsed in the process. Other special precautions for changes in elevation include the following:

- Discuss hazards formally with all performers and personnel prior to any work or rehearsals involving the hazard.
- Inform all performers and personnel of designated safe routes where crossovers and traffic will be routed to avoid going near the hazard. No one should be allowed to approach hazardous changes in elevation unless absolutely necessary.
- Mark changes in elevation on the stage with phosphorescent tape, so that they are more easily seen in the dark or semidark.
- Mark pits, traps, and other hazards with large signs, and barricade them when they are not in use. Mark open pits and traps with a ghost light (standing lamp) when they are not in use.
- Give performers extra rehearsal time to learn to work safely with elevation changes. Do not use children or amateur performers in productions with pits or traps.

[2] "A Survey of Injuries among Broadway Performers," Randolph W. Evans, MD, Richard I. Evans, PhD, Scott Carvajal, MA, & Susan Perry, PhD. *American Journal of Public Health,* 86(1), Jan 1996, pp. 77–80.

LADDERS

OSHA regulations about ladders are also found in both the construction and the general industry standards (at 1926.1053, and 1910.25-26). OSHA also cites an American National Standards Institute standard for ladders (ANSI A 14.4) called *Safety Requirements for Jobmade Ladders.* All ladders used must meet this conformance standard.

Some of the commonsense rules about ladder use include:

- Inspect ladders before use. Destroy ladders with defects rather than make homemade repairs that may not meet standards.
- Never substitute a chair or box for a small ladder.
- Use stepladders only in completely open and braced positions.
- Never stand (or sit) on the tops of ladders. Use only the side with steps for climbing or support.
- Never leave tools on a ladder, and never drop or pitch tools to another worker.
- Use barricades to keep activities or traffic away from ladders when they are in use.
- Wooden ladders are preferred, because they are nonconducting, heavier, and usually more stable.
- Nonskid feet should be installed on straight ladders. Tie off ladders, block them, or have an assistant support them against slipping when you use them.
- For large A-frame ladders with extensions, make sure at least two people assist at the base to secure them against tipping.
- Use straight ladders only on clean, level surfaces. Lean them against a surface at a distance of approximately one-quarter of the length of the ladder. Never lean ladders against free-hanging pipes or other unstable surfaces.
- Rather than carrying heavy objects up a ladder, it is safer to climb the ladder, drop a line, and haul the object up.

ELECTRICAL SAFETY

Workers can avoid dangerous power tools by purchasing only power equipment that is either grounded or double-insulated. They are easy to identify:

- A grounded tool has a three-conductor cord with a three-pronged plug that must be plugged into a grounded outlet. If the tool electrically malfunctions, the ground wire provides a low-resistance path to carry electricity away from the user.

- A double-insulated tool has a two-conductor cord and a special insulation system that does not require grounding. Instead, it has an insulating sleeve on the motor armature shaft, between the motor field and housing, and insulated primary and secondary handles. These tools should have a label or a symbol on them indicating that they are double-insulated.

The National Electric Code (NEC)

The source of electrical standards is the National Electric Code (see Bibliography under "National Fire Protection Association"). OSHA has adopted most of the NEC by reference. Most cities commonly adopt the NEC as their local code. Some communities, however, have their own codes. For example, New York City and Chicago's codes differ somewhat from the NEC, and in some cases they are stricter. Lighting technicians and electricians should check with local authorities to obtain information on local codes.

Some good rules to follow:

- Never clip off ground pins on three-wire appliances or use two-wire adapters to wed incompatible equipment.
- Never use substandard two-wire household appliances, lamps, hair dryers, and power bars.
- Keep electrical service and breaker panels accessible. These electrical panels should have thirty-six inches of clearance in front of them and a three-foot-wide aisle leading to them (29 CFR 1910.303(g)). It helps to paint a yellow line around the area that must remain clear to keep storage from occurring there.
- Install ground fault circuit interrupters (GFCI) on any outlet in damp or wet locations (e.g., ten feet from sinks), as required by (304(f)(5)(v)(C)(5). Also install GFCIs on outlets frequently used for power tools, as required by the construction standards (29 CFR 1926.404(b)(1)). Use extension cords that have GFCIs built into them.
- Discard old lighting instruments with asbestos wire insulation (see chapter 20, Asbestos).
- Make sure there is a completely dead (nonconducting) front on dimmers and light boards.
- Coil temporary wiring neatly, keep flexible cable out of traffic areas, and cover wires that cross walkways with treadles.

LIGHTING

By definition, theatrical lighting is temporary wiring. All codes and OSHA regulations applicable to temporary wiring apply to theater lighting. Here are some especially important general recommendations regarding theatrical lighting:

- Hold safety-and-strategy meetings at the beginning of each work period for the entire crew. Be sure new people know the location of the master switch for stage lighting equipment. If students or trainees are present, review all safety procedures.
- Make sure that each individual knows precisely what his or her responsibilities are as defined by a job description, in the *Handbook of Theatrical Apprentices* (see Bibliography), or some other suitable set of formal guidelines.
- Establish firm lines of command and discipline.
- Arrange work schedules so that no other activities take place on stage while lights are being hung or focused.
- Permit only authorized and trained personnel to work on lighting. Lighting trainees and students should be required to complete courses on stagecraft and lighting, be required to have stage experience, and should be supervised closely.
- Training for lighting personnel should include basic first aid and emergency treatment for shock victims.
- Never work alone on hazardous lighting procedures, such as hooking up panels. Organize buddy system work schedules.
- Before hanging lights, crew members should make sure that there is nothing on their persons that could fall to stage level. They should empty their pockets and remove glasses, jewelry, hair ornaments, and the like or secure them against falling. Secure all tools to workers with safety lines.
- Wear protective clothing appropriate to the work. For example, wear rubber-soled shoes and heat-resistant gloves. Hard hats should be worn on occasions when overhead hazards are present.
- Report immediately any irregularities, defective equipment, or incidence of electrical shock (no matter how slight).
- Know your instruments. Read lighting equipment directions and product information carefully. For example, a quartz lamp can explode when it is used if you fail to follow instructions to avoid touching it with your bare hands at all times.

- Never overload dimmer boards.
- Avoid "spaghetti." Coil cables and cords neatly to avoid safety hazards, heat buildup, and magnetic field interference.
- Have lighting inspected and approved by a licensed electrician.

THE STRIKE

Strikes should receive special attention, because they combine all the previously mentioned hazardous activities and more. Lighting, rigging, and electrical safety precautions are especially important during strikes. Discipline and planning should replace hurried activity. A strike is not a time for a party, nor should you hurry a strike in order to get to a party.

Teachers of theater arts must be especially careful to maintain discipline. Without discipline, students'—especially high school and college students—destructive urges can quickly transform disassembly into demolition.

OTHER SAFETY HAZARDS

This chapter has covered some of the more common hazards. However, a vast variety of other hazards can exist in theaters, such as revolving platforms and elevators. It is incumbent on employers and teachers to research the hazards of any new project or element and make sure the precautions taken are consistent with federal OSHA laws and local codes.

19 Location Hazards

Any location can harbor biological, chemical, structural, and safety hazards. Working in old armories, piers, warehouses, and other venues large enough to accommodate filming, set construction, and shops can also be hazardous. Even well-maintained public buildings and schools can be unsafe locations under certain conditions.

But the worst conditions I have seen are on locations that are chosen to represent a city's underbelly, such as abandoned buildings, shooting galleries, and closed factories. City officials, in their eagerness to bring the film business to their area, often lease these abandoned properties at little or no cost. And they usually do not discuss any potential hazards on the site with the locations scouts.

It has also been my experience that most locations scouts and managers are not trained to recognize the difference between sites that *look* dangerous and those that *are* dangerous. They need to know how to obtain and read an EPA environmental audit, to check certificates of occupancy, to know when structural engineers should be called in, when that flaking paint or falling insulation should be tested.

Until better training is given to locations scouts, it remains for workers to be aware enough to alert their unions or the appropriate governmental agencies when conditions need investigation. Workers in Local 829, United Scenic Artists, can call their union to request an inspection by an industrial hygienist. This approach also prevents reprisals by the employer, since the person who is complaining will not be identified.

COMMON HAZARDS

There probably will always be less-than-ideal conditions on locations. However, there are some conditions that are so hazardous that we must intervene.

Structural Hazards

Rotted flooring, crumbling pilings, rickety staircases, sagging beams, and similar defects are both picturesque and hazardous. Whenever these obvious signs of deterioration are present, a structural engineer must be consulted before work begins. Even with no outward signs of deterioration, a structural engineer should be retained to provide written advice regarding any use of a building that involves heavy equipment, heavy storage, or renovation plans calling for alteration in walls, wall openings, beams, or other support structures.

Fire Safety

If there is an overhead sprinkler system, OSHA regulations require that your employer hold a sit-down meeting with all the workers to instruct them on how it works, what the alarms mean, and what you should do if they go off. If this has not happened, try to find out if the sprinkler system is in working order and if it is a dry- or wet-pipe system.

Wet-pipe systems immediately discharge water if someone accidentally knocks off a sprinkler head during work. Dry-pipe systems give you a short time to turn the system off after a head is accidentally damaged to avoid water damage to the set.

If there are no sprinklers, look for handheld, ABC-type extinguishers located, at most, seventy-five feet apart. The tags on the extinguishers should show they have been recently inspected. OSHA also requires that all workers be trained in their use.

Emergency Exits and Escape Routes

Look around you: There must be at least two escape routes from all areas. Exits or exit signs should be visible from all locations. Fire doors and panic bolts must be in good repair and must *never* be chained or locked while you are working in the location. Inadequate escape routes or damaged fire escapes are items that have to be corrected before the location is safe enough for you to begin to work.

OSHA rules require employers to hold a formal meeting to explain the workspace hazards and emergency procedures whenever new employees arrive on site. If employers do not follow these laws, you need to protect yourself. As soon as you arrive at a new location, familiarize yourself with the fire protection system and emergency egress. Find out how to access fire escapes and where everyone should meet after evacuation. Share the information with others, and let the union know that the employer is not providing the required training.

Changes in Elevation

Any elevated platforms or storage areas, shafts, or holes where people could fall more than six feet must be guarded. Standard railings (either permanent or temporary) and covers over holes must be installed *before* you are asked to work in an area. Once the site is open to nonconstruction workers, such as cameramen and actors, changes in elevation of four feet or more must be guardrailed.

Access to changes in elevation, such as ladders and stairs, must also meet OSHA regulations. For example, stairs having four or more risers or that rise more than thirty inches must be equipped with at least one handrail and one stair-rail system along each unprotected side or edge.

Electrical Hazards

Do not ignore flickering or dimming lights, frequently interrupted power, damaged wiring, or other electrical defects. OSHA also requires outlets used for power tools on construction sites and all outlets within ten feet of a source of water to have ground fault circuit interrupters.

Garbage and Toxic Chemicals

Only professional waste handlers can safely remove refuse that contains animal or human excrement, dead animals, rodent nests, discarded needles, used condoms, moldy materials, and similar unsanitary substances. Only toxic waste disposal contractors can legally remove old chemical products and containers, unidentified or unlabeled substances, asbestos and lead paint waste, and other chemicals.

Waste removal services are expensive, but scenic artists or general laborers must not handle unsanitary, infectious, or chemical wastes. Call the authorities or your union if you are asked to do this work or if you suspect that untrained workers are removing waste. These actions are necessary, because serious diseases can be carried by animal waste or moldy and decaying materials (see chapter 15, Mold and Other Biohazards).

Bathroom Facilities

Clean bathroom facilities must be present in sufficient numbers to accommodate the size of the workforce. If not, portable toilets must be rented until bathrooms can be installed. Portable toilets alone are not adequate. A water supply for washing hands and cleaning up also must be available.

Drinking Water

In older buildings, service pipes and plumbing pipes often are made of lead. Lead also may be found in solder used on potable water pipe (lead was banned for this use in 1986), in faucets, and floor-model watercoolers. The only way to know if water is safe to drink in older buildings is to have it tested. If the water has not been tested, or if the test shows the water is above the accepted limit, ask for or bring your own bottled water.

Eating and Drinking Facilities

Two sections of the sanitation rules (1910.141(g)(2) and (g)(4)) prohibit employees from eating or storing food in bathrooms or in areas where toxic substances are used or stored. Eating lunch or drinking beverages must be done in a sanitary room. If such a room is not available, eating must be done off premises. The traditional practice of setting up craft tables of food near painting, woodworking, welding, and other construction activities is against the OSHA regulations and needs to be changed.

Lead Paint

Buildings built before 1980 should be assumed to contain lead paint, unless actual testing shows otherwise. Even well-maintained older buildings may contain painted-over or encapsulated lead paint that can be made airborne if renovation of any painted surface is planned.

The OSHA Lead in Construction Standard forbids sanding and resurfacing, removal, or demolition of any painted surface unless the paint has been professionally tested and shown to be lead-free. If lead is found, only trained lead-abatement contractors can do the work. Scenic artists must never do this work.

Even when renovation is not planned, dust from lead paint still can be hazardous. Testing should be done if visible paint dust or potential sources of paint dust are noted. Two common examples are:

- *Deteriorating and chipping paint.* Paint that leaves a chalky residue on your hands or is chipping and flaking off walls is a source of lead and lead dust. If the chips reach the floor and are walked on, large amounts of lead dust can be created.
- *Friction surfaces.* Painted surfaces like those on window frames or sliding doors create dust when drawn over each other.

Asbestos

Owners or lessors of any building in which people work, whether currently occupied or previously unoccupied or abandoned, must have an asbestos plan. In August 1994, OSHA released its final rule on asbestos (29 CFR 1910.1001 and 1926.1101). This new regulation requires notification about any asbestos materials to all workers using the site, including maintenance personnel, tradesmen, and even general-office occupants. If you see materials that look like asbestos and you have not been formally told whether or not they are asbestos, know that your employer is not following this law.

Only certified asbestos-abatement contractors are allowed to alter or demolish asbestos-containing building materials. Some sources of asbestos are easily recognized. A list of them is included in chapter 20.

Treated Wood

Structural plywood and outdoor wooden items such as fences, decks, piers, playground structures, and picnic tables usually have been treated with restricted-use pesticides (e.g., creosote, CCA, pentachlorophenol). Some of these treated woods are not green in color and cannot be detected visually. Never burn wood that may have been treated. If treated wooden materials must be cut, sanded, demolished, or modified, you will need extra respiratory and personal-protective equipment and dust-collecting woodworking tools.

Pest Control

Some locations will need pest control, while other sites may be hazardous from previous use of pesticides. Signs that there may be a problem include the presence of many live and dead insects, roach-bait stations, mouse and rattraps, and evidence of insect or rodent damage. Licensed pesticide applicators should be consulted about treating sites for control of serious infestations.

Gloves and respiratory protection should be used whenever cutting or sanding old baseboards and other materials on which pesticides were traditionally applied and sprayed.

Stress from Heat or Cold

Outdoor locations are exposed to weather, and indoor locations are not always properly heated and air-conditioned. Just as there are limits for exposure to chemicals in workplace air, there are limits for the amount of heat and cold that workers should endure. Too much heat can cause serious conditions, such as heat exhaustion and heatstroke. Cold can cause hypothermia. In the extreme, both heat and cold exposure can be fatal. If you think the temperatures are excessive, call your union.

The limits for heat and cold exposure, like those for chemicals, are called "Threshold Limit Values"[1] under OSHA regulations. Based on monitoring of temperature (and humidity in the case of heat), workers may be placed on a schedule of frequent rests, during which they can either cool down or warm up. These standards are rather complex, but employers are required to provide monitoring of excessive heat and cold. (See the Bibliography for the data sheet on heat stress for further information.)

Carbon Monoxide Exposure

Indoor locations are very likely to employ lifts and other equipment powered by diesel or propane fuel. This equipment exhausts carbon monoxide in hazardous

[1]OSHA adopted the American Conference of Governmental Industrial Hygienists TLVs for heat and cold.

amounts. OSHA regulations require employers to monitor carbon monoxide levels. If this is not done, there are ways in which workers can use certain types of household carbon monoxide detectors to check the levels. (See the Bibliography for the data sheet on carbon monoxide detectors.)

A good policy, which was developed by Local 829 of the United Scenic Artists, is to require employers to use electric lifts whenever possible. If diesel lifts are needed for very heavy loads or for heights above ninety feet, a device that captures toxic emissions must be placed on their exhausts.

Enclosed or Confined Spaces

OSHA has special rules for work inside a boiler, tank, pipe, box, or any space that is relatively small in size and has no openings for cross-ventilation. OSHA calls these "confined spaces." One reason for the special concern is that life-threatening amounts of gases and vapors can collect in such areas quickly. Painting or using any product containing solvents in small spaces can fill them with high concentrations of toxic vapors in minutes. Call your health and safety officer for advice if you think you may be working in a confined space.

Environmental Hazards

Old factories and commercial buildings may be contaminated with toxic substances. For example, mercury can contaminate buildings in which neon and mercury vapor lights, thermometers, and other instruments were made. Buildings used for metalworking, electroplating, and jewelry-making can be contaminated with highly toxic metals, like chromium, cobalt, nickel, and even arsenic and lead. Old textile factories can be contaminated with highly carcinogenic, organic chemical dyes and their contaminants, including dioxins.

Treat all old building materials from industrial sites as potentially toxic by using respiratory protection when cutting, sanding, or deconstructing them.

SUMMARY

These are a few of the most common hazards found in locations. There are many more. If you see any conditions about which you are concerned, ask. Your life is worth more than any shot.

20 Asbestos

Media coverage of actor Steve McQueen's battle with cancer and his subsequent death will not soon be forgotten. But it is likely that most of us have forgotten the nature of the cancer that killed him—mesothelioma. This cancer is almost invariably caused by inhalation of asbestos fibers.

SOURCES IN THE PAST

The source of the asbestos that triggered Steve McQueen's illness is not known with certainty. One press account blamed asbestos clothing and padding, which he used for fire protection when he was a race driver. As an actor, however, he could have been in contact with many sources of asbestos.

For example, during Steve McQueen's early days in theater, scenic artists were still buying bags of asbestos fiber and throwing them in scene paint to provide texture. Special-effects people used asbestos to simulate falling snow. Until the early 1980s, instant papier-mâché powders, which were often labeled "nontoxic," contained as much as 80 percent asbestos powder.

Actors were exposed to many other unusual sources of asbestos. For example, in 1986, union workers on the set of *As The World Turns* called me. Actors and crew members reported having respiratory problems from heavy dust in the studio from fake sand used in shooting desert scenes. The sand was actually vermiculite, of the type that was sold in big bags for use in gardening and for insulation. The supplier of this vermiculite was W. R. Grace. I sent the union workers documentation showing that W. R. Grace's Montana mine produced vermiculite that was contaminated with significant amounts of an especially toxic form of asbestos called fibrous tremolite. The *World* stopped turning, and they cleaned up the premises.

W. R. Grace shut down their contaminated mine in 1990. Today, they are inundated with lawsuits from workers all around the country who worked with the vermiculite and subsequently have asbestos-related diseases and cancers.

But this story has a sequel. My files indicate that a week after the vermiculite incident, I received another call from *As The World Turns* asking about fuller's earth, a clay that had just been dumped on actors to simulate conditions after an explosion. Clays are inert mineral dusts that should not be inhaled in quantity. And some forms of fuller's earth clays contain a mineral called fibrous attapulgite, which has the same hazards as asbestos.

This use of fuller's earth is still common. In 1998, a union camera worker was heavily and repeatedly overexposed to fuller's earth over a couple of days on a movie location. The fuller's earth was thrown into a large fan to cloud the scene. The worker developed a lung condition (interstitial fibrosis) and filed for workers' compensation.

SOURCES OF ASBESTOS TODAY

There are many sources of asbestos still lurking in building materials and insulation in theaters and locations today. They may include:

- *Asbestos fire curtains.* These curtains are still in many older theaters. If they are in excellent condition, they do not shed many fibers.
- *Insulation around pipes, furnaces, beams.* Insulation can be in the form of paper, cardboard, a powdery material under a cloth covering, cord and rope (e.g., between metal surfaces around equipment doors and openings), blown or foamed onto beams, and in many other forms.
- *Insulation around electrical equipment.* Fibrous-looking insulation is sometimes packed around lighting fixtures or electrical equipment.
- *Wiring.* This includes both the white fuzzy variety and some wires consisting of plastic/rubber sheathing over fibrous asbestos underneath. These are especially common on old lighting instruments.
- *Composition ceiling tiles.* In the past, many major brands of standard ceiling tiles contained asbestos.
- *Acoustic board and tile.* These contained asbestos in the past.
- *Transite™ and other asbestos boards.* Usually mottled and gray in color, these boards were used extensively in the past. They still are used for insulating boiler room walls and similar applications.
- *Old wallboard and plaster.* Asbestos was commonly added to plaster and to a few old types of wallboard.
- *Spackle plaster repair compounds* were made with asbestos until the mid-seventies. Removing old plaster repairs can be risky.
- *Vinyl floor tiles.* Old tiles commonly contained asbestos. Some new tiles also may contain asbestos. They are hazardous when broken,

sanded, or buffed during maintenance. The mastic under the tiles often contains asbestos as well.

- *Roofing felts, tar paper, and caulks.* Old roofs that have become weathered and dry may release asbestos when disturbed.
- *Old paints* sometimes had asbestos fibers added to them. Both exterior and interior paints like this were once marketed. Asbestos was also used to extend and texture scene paints and glues in the past.

HOW MUCH IS TOO MUCH?

We don't want to become paranoid about asbestos, since standing on a busy street corner will expose us to small amounts of fiber from vehicle brake linings. But a NIOSH-OSHA Asbestos Work Group reviewed asbestos studies and concluded that there is no level of asbestos exposure below which there are no clinical effects. Exposure duration studies suggest that exposures as brief as one day to three months can bring on significant disease in some people.

Dramatic evidence of the effects of low-level exposures can be seen in the asbestos-related diseases suffered by members of asbestos workers' households from the workers' clothing or by persons who have lived near asbestos-contaminated areas. These studies also show that once asbestos fibers are present in our homes or workplaces, they will persist.

Asbestos fibers are virtually indestructible and invisible. They can be stirred up on the slightest air currents and can float almost indefinitely where they can be inhaled easily. They will go through the filters of ordinary household and shop vacuum cleaners. It is crucial to keep them out of places in which we live or work.

ASBESTOS-RELATED DISEASES

Inhaled asbestos is especially associated with three diseases:

- Asbestosis (a lung-scarring disease)
- Ordinary lung cancer
- Mesothelioma (a cancer of the lining of the chest and abdomen).

Asbestosis requires considerable exposure to asbestos and is not likely to occur among people in our business. But relatively smaller amounts are needed to increase the risk of lung cancer and mesothelioma. The risk is dose-related, and the greater the exposure, the greater the risk.

Asbestos has also been linked to other malignancies, such as stomach and intestinal cancer.

ASBESTOS REGULATIONS

The OSHA standards for asbestos in general industry (1910.1001(a)(1)) and construction (29 CFR 1926.58 & .1101) both define asbestos-containing materials as "any material containing more than one percent asbestos." Any time asbestos exposure is possible, the employer must have the work done by workers specially trained and certified for this work.

This essentially means that no worker should ever disturb any building material that could be asbestos unless the employer shows them lab reports indicating that the material is not asbestos.

The law also addresses untested building materials. It calls them "presumed asbestos-containing materials" or PACM. The term PACM is applied to any untested thermal insulation, spray-applied and/or trowelled-on surfacing material, including resilient floor-covering materials installed in buildings in 1980 or before. Building owners must treat PACM as asbestos, whether it is or not, until proven otherwise. This includes labeling of all untested insulated pipes. And information about all the PACM and other asbestos building materials must be compiled in a written asbestos management program.

The law further requires that workers must be notified of the presence of asbestos-containing materials. This includes even employees who will work in or adjacent to areas containing asbestos. For example, if any of the pipes in hallways are insulated with asbestos or PACM, this information must be formally presented to people who work in offices whose doors open into these halls.

This means that workers can assume that building managers and/or employers in buildings built before 1980 are violating the law if:

- Pipe insulation is not labeled
- Workers are asked to disturb suspicious-looking materials without being provided test data
- Workers are not notified about which materials in the areas they work are (or are not) asbestos
- Workers are not given access to the written asbestos management program on request

REMOVE, ISOLATE, OR ENCAPSULATE?

When asbestos is found, a decision must be made whether to remove it, to isolate it, or to encapsulate it. Removing asbestos is costly. In some cases, asbestos can be left intact, sealed behind airtight barriers, such as specially lowered ceilings. This solution is not usually preferred, because later damage to the ceiling or running cable in the ceiling may mean exposing the asbestos.

Another method of containing asbestos is to encapsulate it. Workers apply a special sealant that penetrates the asbestos insulation and hardens to hold it in place. Encapsulation usually is much cheaper than removal, and it at least partially preserves asbestos' insulating and fireproofing attributes. How long the encapsulant will remain intact is a matter for conjecture.

INSULATED WIRE

This author still finds old lighting instruments in theaters that have asbestos wires. The whitish, fuzzy wires are readily recognizable. These wires shed fibers even when new. The old wires shed enough to often require professional abatement in the areas in which they have been stored. And they will shed a great amount of fiber if anyone tries to remove them from the instruments.

Whenever I see asbestos wires, I recall a lecture/training that I did for the theater department of a well-known eastern university in 1992. During a discussion, one young lady said she was thrilled when she was accepted at this school. But the first job she was assigned was to remove and replace the asbestos cords on over two hundred instruments. This she did—with no precautions whatever.

Fortunately, three people from the school's safety office were also in attendance. They told the audience that they were unaware there were hazards of this magnitude in the department, and they vowed to correct them. I understand that the young lady has filed an intent to sue the school in case she is unlucky enough to ever develop one of the asbestos-related cancers.

And this same scenario could be occurring today. As late as May of 1998, I saw illegal advice about removing the wires from these instruments in a popular theater magazine. It is really time to settle the question about these wires once and for all.

From the discussion of the laws above, it should be obvious that removal of the wires cannot be done legally except by trained and certified workers in containment enclosures. For this reason, responsible lighting sales and service companies will not replace asbestos cords on these instruments. Instead, there are really only two options:

- Double-bag the old asbestos-wired instruments, and call a toxic waste disposal company. In some cases, the wires are connected to a cap that can be disposed of and the rest of the instrument saved. Lenses and bulbs also may be able to be saved.
- Send the old instruments to a special asbestos-abatement company. This is more expensive, but will allow you to save the old instruments

for period pieces or museum collections. One company that will do this work is:

A & S Environmental
2261 East Fifteenth Street
Los Angeles, California 90021
(213) 623-9443

THEATER CURTAINS

Theater curtains made of asbestos are common in old theaters. If these are in excellent condition and painted, they do not shed fibers and are not a hazard. They become a problem when they need repair or replacement. For example, union workers called me in 1996 to stop an employer from having them cut and shorten an asbestos curtain in a theater that was being used as a movie location. Instead, there are many types of asbestos-substitute curtains that could be installed.

GLOVES

Old and worn asbestos gloves have long been known to release dangerous amounts of fibers. A study done by the NIOSH-OSHA Asbestos Work Group has shown that even brand-new gloves release substantial amounts of fiber into workers' breathing zones under conditions of normal use. Sources of non-asbestos gloves can be found in any safety supply outlet.

OTHER SOURCES OF ASBESTOS

Old papier-mâché, old asbestos-coated welding rods (these are no longer made), and old sources of vermiculite mined in Montana in the United States are also asbestos-contaminated (the dust may contain 3 to 5 percent tremolite asbestos). Old talcum powders also contained asbestos.

ASBESTOS SUBSTITUTES

Workers may also run into other types of insulation, such as fiberglass or synthetic mineral fiber. These human-made fibers are not without hazards. Studies of fiberglass workers and animal tests provide enough data for fiberglass to be listed as a suspected animal or human carcinogen by several agencies. The American Conference of Governmental Industrial Hygienists (ACGIH) has set a standard of 1 fiber/cubic centimeter for the various forms of glass fibers.

Glass fiber is also the cause of eye and respiratory irritation. Asbestos, on the other hand, does not have any obvious symptoms of exposure.

One type of asbestos substitute is called refractory ceramic fiber. This product may be just as hazardous as asbestos. The ACGIH set the same workplace standard for this fiber as for asbestos (0.1 fiber/cubic centimeter). This author has seen stage fireplaces insulated with this fiber in blanket form. In my opinion, refractory ceramic fiber has no place whatever in any production.

21 Fog, Pyro, and Other Effects

Most large productions today involve special effects. These effects have occasionally caused accidents ranging in seriousness from the pyrotechnic effects that set Michael Jackson's slicked hair on fire to the deaths of actor Vic Morrow and two young children during the filming of *The Twilight Zone.*

In addition to accidents, there are health effects associated with special effects. Performers are the only workers in the world who are expected to do creative, highly demanding work in air that is deliberately polluted with chemical fogs and smokes.

HEALTH STUDIES

Three scientific studies have found that high percentages of performers responding to questionnaires claim to have endured health effects from the fogs and smokes:

1. The National Institute for Occupational Safety and Health (NIOSH) Final Report released in August 1994 (HETA 90-355-2449).
2. "Health Effects of Glycol-Based Fog Used in Theatrical Productions." Harry H. Herman Jr., *Report to Actor's Equity Association,* July 1995.
3. Report on a study of about twenty-five Local 802 pit musicians at *Beauty and the Beast* on Broadway by Dr. Jacqueline M. Moline, Mount Sinai–Irving J. Selikoff Center for Environmental and Occupational Medicine, January 17, 1997.

Symptoms they reported included eye and throat irritation, increased incidence of colds and respiratory infections, sore throats, throat infections, headaches, dizziness, fatigue, bronchitis, various types of pneumonia, allergies, and asthma.

The first two of these major studies did not show a connection between the theatrical fog and asthma, but the study which investigated the health complaints of the twenty-five pit musicians at *Beauty and the Beast,* where both fog and pyrotechnics are used, found that:

> . . . analysis of the pulmonary function tests showed that there was a statistically significant decrease in forced vital capacity from pre- to post-performance. . . . Of the fourteen musicians who were present on both screening days, 10 of 14 (71%) showed small airway dysfunction. . . .
>
> . . . A large percentage of the musicians are suffering from symptoms related to the irritative effects of the work environment. Several musicians now require medical care and medication to treat their symptoms which have developed or worsened since taking part in this production.

Many other short studies and articles on health effects associated with fogs and smokes are listed in the Bibliography and Sources (page 209). I hoped to include a study by the Department of Community Medicine at Mount Sinai School of Medicine in New York City, but it has not yet been released (even though it is more than a year overdue). Another study has commenced at the University of British Columbia School of Occupational and Environmental Hygiene. This issue is far from settled.

LAWSUITS AND COMPENSATION

Some performers and technicians seek compensation for damages they claim special effects have done to their health. This author is personally aware of three personal injury lawsuits, four workers' compensation claims, and one suit brought forth by an asthmatic worker claiming that the use of fog did not accommodate her disability. Theaters paid settlements to performers in two of the cases, two workers obtained compensation (one was for asthma), and the outcomes of the two other compensation suits and one lawsuit are still pending at the time of this writing.

Some of these cases involve singers and musicians who have become sensitive to special effects and can no longer perform or are limited to working in those few venues that do not use special effects. These performers' disabilities are no less pressing than those of nurses and doctors who have suffered the effects of exposure to rubber latex. And, just as doctors and nurses have been compensated, performers should be remunerated by those who have sold the hazardous products and by the employers who have insisted on using such products.

EXPOSURE STANDARDS

Advocates of fog effects claim, correctly, that most of the chemicals used today are of relatively low acute toxicity. They fail to appreciate that many supposedly nontoxic chemicals provoke serious allergic responses in some people. People can die from traces of nontoxic proteins in peanuts or natural rubber. When chemicals are added to air—no matter how nontoxic they are to the majority of the population—some people who inhale them will have problems.

In addition to allergy, it is now known that particles of any type that are lodged deep in the lungs cause adverse effects in many people. This is the basis of the Environmental Protection Agency's Air Quality Indexes (AQI) for particulates.

EPA Air Quality Index (AQI) Standards

AQI standards are used by public health services to determine when it is time to alert people with breathing and heart problems to stay indoors and not to exercise. EPA sets two standards for particulates, one for those 2.5 microns in diameter and smaller and another for those 10 microns and smaller. Two categories are necessary because it has been found that smaller particles are more damaging to the respiratory system than larger. Most haze, fog, and smoke-mist particles range between 1 and 5 microns. Some manufacturers say their hazes can be as small as 0.5 microns.

U.S. EPA AIR QUALITY INDEX (AQI)*

	PARTICULATE MATTER (PM)	
AQI DESCRIPTION	$PM_{2.5}$, 24-hr**	PM_{10}, 24-hr**
good	0.015 mg/m^3***	0.05 mg/m^3***
moderate	0.065	0.15
unhealthy for sensitive groups	0.10	0.25
unhealthy	0.15	0.35
very unhealthy	0.25	0.42
hazardous	0.35	0.50
"	0.50	0.60

*64 FR 42529-42573. Aug. 4, 1999.
**Particulates under 2.5 microns and under 10 microns in diameter.
***mg/m^3 = milligrams/cubic meter.

The concentrations in this chart could be used to determine when the amounts of solid particulates from inorganic smoke fumes or pyrotechnic smoke are in excess. It is less clear how applicable the AQIs are to the irritant effects of such tiny liquid particulates as oil and glycol. However, at this time the author thinks the EPA AQIs provide better guidelines than any of the other standards for particulates.

Threshold Limit Values (TLVs)

Advocates of theatrical fog most often refer to the TLVs, which constitute industrial air quality standards of the American Conference of Industrial Hygienists. This author thinks these standards are not applicable, since performers are presently working in ways never anticipated by the standard setters. Singers take huge breaths of air and spin out the inhaled air slowly. Dancers stress their bodies and lungs like athletes. Even slight effects on respiratory systems or vocal cords can affect performance.

In addition, the TLVs are designed to protect only healthy adult workers. Many performers are children. And many performers are not healthy. Some already have lung problems or even serious illnesses such as AIDS. TLVs are not designed to protect these groups.

The American Society of Heating, Refrigerating, and Air-conditioning Engineers suggests applying one-tenth of the TLV for substances that do not have indoor standards. But there are no TLVs for many of the fog chemicals. And if the TLV for nuisance particulates (10 milligrams per cubic meter) is used, the indoor standard for fog and smoke particulates would be 1.0 milligram per cubic meter. This limit cannot be expected to protect people who have asthma or sensitivities to fog or smoke.

FOG AND SMOKE MACHINES

There are many types of fog and smoke-generating machines. For a complete description of how these work, consult *Introduction to Modern Atmospheric Effects,* a booklet produced by Entertainment Services & Technology Association (ESTA). (Go to *www.esta.org* to order the booklet, or see the Bibliography and Sources.)

Some types of machines employ heat. Awareness of temperature is important when one assesses the hazards of fogs and smokes, as heating can decompose the glycols, oils, or other fluids and thus release many toxic chemicals. Overheating machines will decompose the fluids in amounts large enough to be immediately noticed by their strong odor. But it is also possible that even properly operating machines release some potentially hazardous decomposition products.

ESTA says that the pump-propelled glycol fog systems heat fog fluids to usually less than 644 degrees Fahrenheit (340 degrees Centigrade). This temperature is hot enough to provoke decomposition of some fluids. The booklet also says that the gas-propelled machines heat the fluids to their boiling points. But most of the fluids do not have a boiling point. Instead, they have a boiling range of temperatures because the fluids are mixtures of water and glycols, mixtures of different glycols, or mineral oils, which are composed of various hydrocarbons. In these cases some components of the fluids may decompose into more toxic chemicals at boiling range temperatures.

TYPES OF EFFECTS

There are many types of fog and smoke. They function differently on stage, and each is made differently.

Haze

This effect is usually imperceptible to people and only shows itself when beams of light are reflected against the particles. Haze particles are extremely small in size, usually less than one micron in diameter. Their miniscule proportions will enable them to be inhaled into the deepest parts of the performers' lungs. As they must be made of substances which evaporate very slowly, hazes are usually composed of mineral oils, polyethylene or polypropylene glycols (low molecular weight plastic polymers), and triethylene glycols.

The biological effect of the haze particulates in the lungs' air sacs has never been addressed by the industry. But it is possible that these particles evaporate so slowly that they may dwell in the alveoli for long periods of time, perhaps indefinitely.

Fog

This effect utilizes larger droplets, usually 2 to 5 microns in diameter. The drops appear to drift and may evaporate either quickly or slowly. They are most likely to be composed of water/glycol mixtures or mineral oil.

Low-Lying Fogs

These fogs hug the floor because the air in which the mist droplets are suspended is colder and denser than the original room air. Water and very cold substances such as dry ice or liquid nitrogen are most often used to create the fog. However, any liquid inert gas can be used. Some of the low-lying fogs also employ chemical fog fluids in their systems.

The fogs generated by cryogenic gases and water constitute water mists. But the air around the mist droplets is rich in the cryogenic gas. Nitrogen-rich air can reduce the amount of oxygen needed to breathe. Carbon dioxide can do this as well, but is also toxic and can cause unconsciousness in a few minutes at levels above 7 percent. This was demonstrated recently at the

San Francisco Opera during a performance of *Electra*. A woman playing the role of Clytemnestra was directed to play dead on the floor next to a pool from which dry ice fog was emanating. She had a seizure on stage from overexposure to the gas.

Because low-lying fogs evaporate quickly, large amounts of fluid and cryogenic gas may be needed to maintain the effect's presence. If the fluid is water, the primary concern is the amount of inert gas released. One manufacturer, Praxair, has addressed this problem by creating fogs with liquid nitrogen and liquid oxygen, which are mixed in the same proportion as in fresh air. The dissipation of this type of fog will actually improve air quality.

Mists

These spectacles simulate such localized effects as storm scenes and mists around a waterfall. Since mists fall out of the air, they must be constantly replenished as the surface underneath becomes wet. These mists may be composed or water or other chemicals.

FOG CHEMICALS

The majority of commercial fogs are made of one or more of the following chemicals: water, glycols or glycerin, and various mineral oils.

Water

Fog machines producing water mists are now available. Some permanent entertainment facilities use steam and ultrasonic mist generators.

Some of these systems create the same hazards as humidifiers. For example, unless distilled water is used, ultrasonic machines will create a fine, irritating dust composed of the minerals and microorganisms present in ordinary tap water. Fog systems employing boiler steam are likely to contain rust inhibitors, antioxidants, and other toxic boiler water additives, despite the filters or traps which are intended to prevent these chemicals from getting into the steam.

Cold water mists can also be hazardous if water is contaminated or allowed to stand long enough to sustain the growth of such microorganisms as those causing Legionnaires' disease. Makers of some water mist machines claim they filter the water—but these filters usually capture only those particles large enough to clog machine nozzles, while microorganisms pass easily through them.

Some water-glycol fog products may be sterile at first and later become contaminated with microorganisms. This occurrence is especially likely in fluids containing such compounds as propylene glycol, which can serve as food for some microorganisms. It is very important to clean out these machines frequently.

Fog machines that supply the water from a reservoir in the machine may have a biocide in the water to control growth of microorganisms. This biocide, however, is also a potential source of toxicity or allergy for some people.

Clean, fresh, water mists are very safe, but the resulting high humidity may be uncomfortable and could affect hot lights and electrical equipment. It is also difficult to control water mist fog density.

Glycols and Glycerin

These fog chemicals are hygroscopic—that is, they attract and organize water molecules around themselves. Such an accumulation causes irritation of the skin, eyes, and respiratory tract and may also create a dryness in some performers' throats. This final reaction is a transient effect, but singers or actors who work night after night with a dry, irritated throat can do permanent damage to their vocal cords and careers. Chemicals of this type include:

- propylene glycol
- diethylene glycol
- butylene glycol (usually 1,3-butylene glycol)
- triethylene glycol
- dipropylene glycol (oxybis propanol)
- glycerin (glycerol)
- polyethylene glycol (E 200, made by Dow Chemical)
- polypropylene glycol
- ethylene glycol (classic antifreeze, occasionally used)

Some of the glycols also cause allergic reactions. And at high levels, glycerin and most of the glycols have toxic effects. Triethylene glycol may even endanger a person's reproductive faculties. Glycerin and most of the other glycols have never been tested for possible links to cancer, birth defects, or reproductive system damage. The levels at which these toxic effects occur are not expected to be reached if fog and smoke generation is done properly.

The polypropylene and polyethylene glycols are of particular concern because they are unlikely to evaporate at all, are very inert, and are the particulates that are most likely to remain in the lung's alveoli indefinitely.

Oils

In the 1980s, Smith and Wesson made a fog product called Pepper Fog7, which was actually Number Two fuel oil! This mist is also a fire hazard in high concentrations.

Over the years, almost every type of oil has been used. Even today, it is not uncommon for special effects technicians to supplement oil fog fluids with baby oil or laxatives from a local pharmacy. In fact, this form of substitution is what allegedly occurred in one of the ongoing personal injury lawsuits involving special effects. Oils used, past and present, in special effects include:

- mineral oils ranging from cosmetic, medicinal, and food grades (e.g. baby oil) to industrial grades (e.g. cutting oils)
- kerosene and various grades of fuel oil
- diesel oils (for outside use)
- paraffin oils (for outside use)
- vegetable oils

Most fog manufacturers use only highly refined mineral oils of various molecular weights. These oils are nonirritating to the skin but can be damaging in other ways. For example, baby oil (a refined mineral oil) is soothing to the skin, but it causes pain and redness when gotten in the eye. This effect occurs when the surface tension of the liquid in the eye is altered. This effect may also occur in the sensitive membranes in the upper respiratory system. This is probably the reason that the major NIOSH study of performers exposed to fog effects recorded the same complaints from performers, whether they were working in glycol or in oil-fogged shows.

Inhalation of large amounts of any grade of mineral oil can cause life-threatening chemical pneumonia. Mineral oil inhaled in significant quantities is also known to remain in the lungs for years, thus causing complications. However, there have been no studies made to determine the fate of a performer who inhales small amounts of mineral oil mist night after night—whether tiny amounts remain in the alveoli is unknown.

Vegetable oils have been used as well. They were touted as safer than mineral oils because of their natural origins. However, OSHA makes it clear that vegetable "oil mist presents the same safety and health hazards as all of the physical irritants."[1] Vegetable oil mists are flammable in high concentrations. There may also be an increased potential for one to suffer allergies to vegetable oils. But at this point no one knows, since there is even less data available on vegetable oil than on mineral oil exposure.

[1]Federal Register: 57 FR 26323 (1992).

OTHER ATMOSPHERIC EFFECTS

Inorganic Chloride Fumes

Most of these products were developed originally to study airflow patterns in ventilation systems and for other, similar purposes. They were never intended for theatrical purposes, and their MSDSs usually contain strong warnings about inhaling the smoke. One manufacturer's original MSDS recommends that users employ self-contained breathing apparatuses under conditions of heavy exposure! Nevertheless, these products still are being used by some special effects people. The chemicals in these products may include:

- ammonium chloride
- titanium tetrachloride
- zinc chloride
- calcium chloride, tin chloride, and other chlorides.

These can be used in small quantities to make a haze or in large amounts to simulate visible fog. They are fumed by methods such as putting smoke cookies in braziers, lighting smoke candles, and setting off bombs. The smoke from ammonium chloride will contain small amounts of hydrochloric acid and ammonia. Titanium tetrachloride is too irritating for effects other than those used outdoors.

Real Smoke

Smoke can be created by burning, on stage, such organic substances as:

- frankincense (olibanum gum, Bee Smokers)
- rosin
- charcoal
- cigarettes and cigars (burned in large amounts)
- paper
- naphthalene and anthracene (used outdoors)
- rubber tires (used outdoors)

Toxic smoke is emitted by all natural organic substances. It matters little if you burn wax, coal, oil, wood, incense, hamburger, or tobacco. Just because we like the smell of incense or burning autumn leaves does not make their smoke healthier.

In fact, leaf burning is now prohibited in most urban locations. There are a number of studies showing that burning leaves in the fall generates large quantities of carbon monoxide, particulates, and at least seven proven carcinogens. Leaf-burning pollution created breathing problems in a majority of asthmatics.[2]

Black smoke needed for outdoor movie scenes may be from burning naphthalene, anthracene, or even old tires. These smokes contain many highly toxic substances including cancer-causing benzene and polycyclic aromatic hydrocarbons.

Dusts and Powders

These are often used to simulate conditions such as after an earthquake, explosion, or dust storm. This is very hazardous for those exposed to the dust. Examples include:

Mineral dusts such as fuller's earth, vermiculite, and talc are all toxic mineral dusts. Some of these products contain respirable silica which can cause lung damage, asbestos (tremolite), or other fibrous minerals (attapulgite) which are associated with asbestos-like diseases. In 1998, a union camera worker was overexposed repeatedly over only a couple of days on a movie location to fuller's earth and developed a partially disabling lung condition for which workers' compensation was obtained. (See chapter 20 on asbestos and other hazardous mineral dusts.)

Organic dusts such as wood dust, wheat and other grain flours, and starch have also been used. All organic dusts are explosive in high concentrations. To illustrate, during a filming of an episode of *Evening Shade,* Burt Reynolds used a prop fire extinguisher which was accidentally charged with cornstarch baby powder instead of talcum. Using it to put out a fire resulted in a fireball that singed the star and destroyed his toupee.

The organic dusts are also very likely to cause allergies. Baker's asthma is an occupational disease that affects some people who work with flour. Wood dust should never be used since it is sensitizing, toxic, and regulated by OSHA.

Natural and Synthetic Fibers

All fibers, large and small, natural or synthetic, can cause lung diseases. Years ago, lung-damaging asbestos was used to simulate snow. But other fibers such as cotton, jute, and hemp have all been associated with lung diseases in textile workers. Synthetic nylon flocking fibers are now known to cause a disease called flock workers disease. Recently, polyester fibers also were associated with lung problems.

---

[2]A number of publications were summarized in an article in the *American Journal of Public Health,* 84(10), Oct. 1994, p. 1696 which made these points about leaf burning.

A thirty-seven-year-old nonsmoking special-effects coordinator in the moviemaking industry was exposed for only two days to artificial snow composed of polyethylene fibers. Over the next eleven weeks he developed a serious, sleep-depriving cough and postnasal condition. During those eleven weeks, doctors had tried everything including surgery to reconfigure his nasal septum and to remove a turbinate bone. Only after a painful washing of the bronchi did the disease begin to resolve. The biopsies on the whitish plaques in his lungs revealed polyethylene fibers.[3]

There are new snow machines that use soap and detergent foam. These are presumed to be safer because the soap particles are too large to be inhaled deep into the lungs and they dissipate or melt quickly when they land on something.

MISCELLANEOUS CHEMICALS

Pigments and Dyes

These can be used to color smokes and effects. They should be avoided if possible and lighting used to provide color. Many are toxic and many people are allergic to them.

Fragrances

Approximately 10 percent of the adult population in the United States is allergic to one or more fragrances. Adding fragrances to fogs increases greatly the chance of people reacting adversely to them.

PYROTECHNIC EFFECTS

Theatrical pyrotechnics are potentially capable of creating ear-damaging sound, eye-damaging light, and airborne toxic chemicals. While damage to the ears and eyes can be dramatic and obvious, potential health problems from inhalation of the smoke are not so easily addressed.

Pyro Chemicals

Hundreds of chemicals can be used in indoor and outdoor pyrotechnics. Each chemical has one or more functions in the chemical reactions that occur on combustion. Most common pyrotechnic mixtures consist of an oxidizer, a fuel, a source of carbon, and various additives such as chlorine donors to enhance color and other chemicals to modify appearance or sound.

---

[3]Letter to the Editor in *The New England Journal of Medicine* (3/18/99, 340(11)).

Emissions

Compounds such as oxides and chlorides of almost any metal in the periodic table can be present in pyrotechnic emissions. Indoor pyrotechnics should not use the most toxic metals such as arsenic and lead, but outdoor effects commonly contain lead and other toxic metals. In fact, some movie bomb effects are made primarily of lead azide and emit large amounts of lead fume.

The metals most often found in dust left from pyrotechnics include compounds of potassium, titanium, sulfur, calcium, iron, and aluminum. These are alkaline in nature and irritate the respiratory system. Silicon oxides are often present as well.

In addition to metal compounds, pyrotechnics emit sulfur oxides, nitrogen oxides, chlorine and hydrochloric acid and many other irritating gases. Carbon monoxide and carbon dioxide are also produced.

HEALTH EFFECTS FROM PYROTECHNIC SMOKE

1. Respiratory and eye irritation can occur from exposure to pyrotechnic smoke. Many of the solid particulates and gases are irritating and some are outright corrosive such as many of the chlorine-, nitrogen-, and sulfur-containing acid gases.
2. Acute and chronic allergic responses of the respiratory system, eyes, and skin can occur from exposure to pyrotechnic smoke. Many of the solids and gases are sensitizing, including those containing chromium compounds and the sulfur oxides. The dyes used to color smoke also may be sensitizers.
3. Acute and chronic systemic toxicity are associated with exposure to small amounts of lead, arsenic, barium, mercury, antimony, and some other metals. Chronic toxicity affecting the nervous system can be caused by some metals including lead, mercury, and manganese.
4. Cancer is associated with many of the metals, chlorine compounds, and organic chemical dyes.
5. Long-term lung problems are associated with the inhalation of fine dusts of silica, aluminum oxide, and other inert compounds.

OTHER SPECIAL EFFECTS

Use of firearms, flash pots, open flames (such as real fireplaces and flaming torches), lasers, strobe lights, and a host of other special effects all require very specialized knowledge and preparation. Be sure personnel have the appropriate experience and hold any certificates or licenses necessary to use the effects.

RULES FOR USING SPECIAL EFFECTS

1. All designers of special effects must appreciate that they literally take the health and breath of all of the performers, crew, and even the audience in their hands when they decide to put anything other than air in the air! Consider every use of airborne special effects carefully. Never dismiss any individual's claim of health effects lightly.
2. If pyrotechnics are used, the pyrotechnician must be qualified under the requirements of the particular state or city (e.g. New York City has their own requirements). All necessary permits and licenses for the pyrotechnic, fire, or laser effects on stage must be in order. Abide by all the regulations for storing, transporting, and handling of hazardous effects.
3. The precautions taken with pyrotechnics should be consistent with those in the Western Fire Chiefs Association's Article 78, "Special Effects in Motion Picture, Television, and Theatrical and Group Entertainment Productions" and the National Fire Protection Associations standard: NFPA 1126 Code for the Use of Pyrotechnics Before a Proximate Audience. The precautions taken with fire effects should be consistent with those in the Western Fire Chiefs Association's Article 77 and NFPA 160 (see Bibliography and Sources).
4. Insurance must be sufficient to cover any damages which might occur from accidents or fires related to the use of pyrotechnics.
5. Never use pyrotechnics, fog, haze, or any other special effect for which the ingredients are trade secrets or unknown. It is one of the insanities in our businesses that performers are told not to take a job if they are allergic to the special effects, and then the identity of the ingredients in the special effects is withheld from them. In this case, they can't possibly know if they are allergic to the effect or not!
6. Purchase only products whose MSDSs reflect the actual use of the product. For example, MSDSs that only cover the hazards of the unaltered fog fluid ingredients rather than the hazards of inhalation of the ingredients in mist form are not proper.
7. Select the safest product for the job. If children, elderly, ill, or other high risk individuals will be exposed, use only nonchemical cryogenic gas or water methods.
8. Consider the limitations of the theater or facility before electing to use fog, smoke, or pyrotechnics. For example, if the ventilation system is not capable of keeping chemicals or smoke from pyrotechnics

from reaching the audience, use only cryogenic gas methods or water. And stage ventilation should be capable of completely removing fog and smoke between acts.

9. Plan to use fog or smoke in low, safe concentrations. Avoid billowing, thick effects which would endanger actors' health and safety, obscure exit signs, etc.
10. As required by OSHA (see chapter 2), hold hazard communication training for all personnel who may be exposed to the fog or smoke before you use it, provide access to MSDSs, and inform them about potential hazards. Other hazards such as safety hazards from obscured vision or slippery floors should also be discussed. If there are people under 18 years of age in the cast, include their parents in the meeting. Young children should not be exposed to chemical effects or pyrotechnic smoke.
11. Inform audiences at the box office and in programs about all the special effects that will be used. Fog and smoke poses a potential hazard to people with lung problems or allergies. People with heart conditions need to know that pyrotechnics or gunshots will be used. Strobe lights can trigger seizures in some people with epilepsy.
12. Set up lines of communication for health or safety complaints should there be adverse reactions or problems after the material is in use.

22 Theatrical Makeup

Throughout history, men and women have sacrificed health for cosmetic effects. Women in the court of Queen Elizabeth I persisted in wearing white lead paint (ceruse) on their faces, even though they knew it ruined their skin and made their hair fall out. In the eighteenth century, one well-known actress died from using lead-laden makeup.

LEAD AND MERCURY

Today, acutely toxic chemicals like lead and mercury are still found in foreign cosmetics—even some sold in the United States. For example, kohl, a mascara made of lead sulfide and antimony sulfide, has been used for centuries to make up children's eyes in the Middle East, India, Pakistan, and some parts of Africa. One U.S. health department found that use of kohl caused high blood-lead levels in eight children. Two of the children's mothers purchased the kohl in this country.[1]

For another example, several Mexican-made, mercury-containing, beauty creams are also used in this country. One is known to have caused elevated mercury levels in 104 people. Because mercury can penetrate the skin so easily, elevated mercury levels were also found in some persons who never used the cream, but were close household contacts of cream users![2]

Mercury is still allowed in cosmetics by the FDA in eye-area products in very low concentrations (0.0065 percent) in order to prevent serious eye infections in users. Mercury preservatives in these very small amounts may cause allergies in a few people, but they are not enough to cause toxicity.

[1] *American Journal of Public Health,* Vol. 86, No. 4., April 1996, pp. 587–588.

[2] *Physicians' Bulletin,* San Diego Dept. of Health Services, May 1996, Press Release # 31-96, CA Dept. of Health Services, Sacramento, *The Mortality & Morbidity Weekly Report* (CDC) 45(19), May 17, 1996, pp. 400–3, and *Ibid,* 45(29), July 26, 1996, pp. 633–5.

These incidents are reported here to emphasize the importance of using only FDA ingredient-labeled makeup.

HAZARDS TO WEARERS

Despite the FDA's regulations, some individuals will have reactions or allergies to makeup. Both ordinary consumers and performers are at risk. There are numerous documented incidents of makeup affecting individual actors adversely. One well-known example is Buddy Ebsen's serious reaction to a shiny aluminum makeup, which cost him the role of the Tin Man in *The Wizard of Oz*.

HAZARDS TO MAKEUP ARTISTS

A number of studies show that beauticians and cosmetologists suffer a higher incidence than average of lung problems, like asthma and chronic bronchitis, more skin rashes, and more frequent kidney and liver damage. Some studies also show that they have a higher incidence of cancer and reproductive problems, like toxemia of pregnancy and miscarriages.

No similar studies have been made of diseases in theatrical makeup artists, but it is clear that they are exposed to some of the same chemical products. It is important, then, to understand the nature of these substances.

HOW WE ARE EXPOSED

In order to harm you, makeup and beauty products must enter your body. They must do this by one of three routes of entry: skin contact, inhalation, or ingestion.

Skin Contact

Some makeup chemicals can cause skin diseases, such as irritation, infection, and allergic reactions. Some makeup chemicals, hair dyes, and solvents also can penetrate the skin and enter the bloodstream.

Inhalation

Inhalation of powders, aerosol sprays, and airbrush mists is another way makeup and cosmetic chemicals can enter your body. Studies have shown that inhalation of aerosol hair sprays can damage or destroy the tiny hairlike cilia that sweep foreign particles from the lungs. When the lung's defenses are weakened in this way, inhaled substances can cause even more damage.

The smaller the particles of dust or mist, the deeper they can penetrate the lung. In the deepest part of the lung, the air sacs (alveoli) are the most vulnerable. Face powder particles and airbrush mists are examples of small particles that can be deposited deep in the lungs. Studies show that tiny particles of inert minerals, such as those used in cosmetics (e.g., talc and kaolin) can remain in the alveoli indefinitely.

Ingestion

Ingestion of lipsticks, wetting brushes with the mouth, and eating, smoking, or drinking while applying makeup can put cosmetic ingredients directly into your digestive tract. Cosmetics are also ingested when the cilia in the upper portion of the lungs raise mucous and dust particles up to the back of the throat, where they are swallowed.

SKIN DISEASES

Acne

The most common reaction to cosmetics is an infectious reaction of the skin. Especially common is a condition known as "acne cosmetica," or cosmetic acne. (Cosmetic acne should not be confused with "acne vulgaris," which is associated with the onset of puberty.) Cosmetic acne is usually a mild condition. Small pimples appear and disappear intermittently and affect women from their twenties through their fifties.

Other types of acne and skin infections can result if cosmetics support bacterial growth or irritate the skin.

Allergies

Many people develop allergies to chemicals in cosmetics. It is estimated that one person in ten is allergic to fragrances in cosmetics. Some of the preservatives and humectants (e.g., propylene glycol) also cause allergic responses in a few people.

Chrome and nickel compounds have been known to cause severe allergies and skin ulcers in industrial workers exposed to them. Chrome compounds can be found in some eye cosmetics, especially in blue and green hues. And while nickel should not be used in cosmetics, nickel allergy has been documented in hairdressers.

In fact, the percentage of the general U.S population that are allergic to nickel has risen in the last few years from 10 to over 14 percent. Experts think that intimate contact with nickel in metal alloys used in earrings and body-piercing jewelry is the reason.[3]

The greatest potential for serious allergic reactions to theatrical cosmetics is in our use of natural rubber latex and foam products, such as eyelash adhesives, special effects makeups, and face-molding compounds.

Allergies to natural rubber are well-known through the experience of doctors and other medical workers who wear latex gloves daily. Somewhere between 10 and 17 percent of medical professionals have developed the allergy.

---

[3]*New York Times*, "When Body Piercing Causes Body Rash," Denise Grady, Tuesday, October 20, 1998, p. F8.

Symptoms may include skin rash and inflammation, hives, respiratory irritation, asthma, and systemic anaphylactic shock. Between 1988 and 1992, the FDA received reports of a thousand systemic shock reactions to latex. As of June 1996, twenty-eight latex-related deaths had been reported to the FDA.

While there are no systematic studies of special-effects latex makeup allergies, this author can testify that her eyes have swelled completely shut on application of rubber latex eyelash adhesive. Many other people have reported similar experiences. Fortunately, there are synthetic substitutes for almost all natural-rubber makeup products and gloves.

Irritation

Chemicals that are caustic, acid, or strong oxidizers can harm the skin by attacking its surface. Examples include sodium and potassium hydroxides, which can be found in cuticle softeners and hair relaxers and removers. An example of a strong oxidizer is peroxide, which is used to lighten facial hair. Organic solvents, such as alcohol and acetone, can also irritate the skin or dry it out by removing natural oils.

Cancer

Sunlight is the major cause of skin cancer, and both natural and tanning-salon light can cause cancer. Some chemicals have been shown to cause it, too. One example is old-fashioned carbon black, which was common in mascara until it was banned for use in cosmetics by the U.S. Food and Drug Administration (FDA).

Many cancer-causing and highly toxic pigments are approved for use in artists' paints and materials. Some may even be labeled "nontoxic," because used as directed, there should be no significant exposure. Using these products directly on the skin, however, is not a directed use and is not advised!

EYE DISEASES

The skin around the eyes is more sensitive and more easily penetrated than facial skin. All types of skin diseases (infection, irritation, allergies, and cancer) that affect facial skin also can affect the skin around the eyes. The membrane covering the eye and lining the eyelids (the conjunctiva) can be affected by cosmetic chemicals, producing inflammation (conjunctivitis).

Scratching the eyeball during application of eye makeup is the most common eye injury related to cosmetics. Once an eye abrasion has occurred, the possibility of infection increases. The most important thing to remember about these infections is that they proceed with extreme rapidity, and immediate treatment for all painful scratches is recommended. Although most scratches from mascara brushes do not result in infections, those that do can cause ulcers on the cornea, clouding of the cornea, and, in rare cases, blindness.

INFECTIONS TRANSMITTED

Makeups can provide a hospitable environment for many microorganisms. The preservatives in makeup are added to increase shelf life, but they cannot prevent an infectious organism from being transmitted from one person to another.

Examples of just a few of the microorganisms that could survive on makeup include cold viruses; bacteria, such as staphylococcus, streptococcus, and impetigo; fungal infections; and highly infectious viruses such as hepatitis A and herpes simplex. Hepatitis A in particular can remain active for months, even on dry surfaces. The AIDS virus probably cannot be transmitted by makeup. But makeup and all personal items such as razors, nail-care tools, and similar personal grooming items that might draw blood or contact acne or open sores should not be shared.

LABEL-READING TIPS

Consumer Makeup

These labels are required to carry a complete list of ingredients. An exception is occasionally granted by the FDA to certain manufacturers who claim that certain of their ingredients are trade secrets. These products can be identified when the phrase "and other ingredients" is included on the label. Trade secret ingredients should not be used by people with skin allergies, since it may not be possible to find out what is in the makeup.

Professional Makeup

Professional-makeup manufacturers are exempt from complete ingredient labeling requirements. Many, however, list their ingredients anyway. These are the products that should be preferred, both for the ability to identify ingredients that may be causing symptoms and to choose products best suited for the intended theatrical effect.

"Not Tested on Animals"

The FDA requires testing of all cosmetic ingredients. Products that claim not to be tested are only claiming that the product as a whole has not been tested. But to market legally in the United States, they must have purchased their ingredients from sources that provide tested and certified cosmetic-grade chemicals.

Natural Ingredients

There is absolutely no reason to trust a natural ingredient more than a synthetic one. Somehow, we have forgotten the millions of years and the millions of deaths it took mankind to distinguish between the mushrooms and the toadstools, the henna and the hemlock. We need to be reminded that the deadly allergies to rubber are caused only by *natural* rubber, because proteins from the sap of the rubber tree are present. And if we trust natural minerals mined from the earth, we need to be reminded that asbestos is a natural mineral.

Both synthetic and natural ingredients can be hazardous. Be suspicious of any product label that induces you to prejudge the product's safety by its natural origins. Instead, you want the ingredients to be certified, tested, cosmetic-grade ingredients that are disclosed by name on the label.

Use As Directed

The FDA has different safety standards for makeup ingredients intended for use around the eyes, for lipsticks, and for face makeup. For example, lipstick ingredients must be tested for ingestion hazards. The eye makeups must protect against infection and be suitable for the especially thin and sensitive skin around the eyes. Makeups, therefore, are considered safe only when used as directed.

COSMETIC INGREDIENTS

Most makeup ingredients fall into a few basic categories: minerals, such as talc and kaolin; vegetable powders, such as cornstarch; oils, fats, and waxes; pigments and dyes; and preservatives.

Minerals

Face powders, makeups, and rouges are likely to contain minerals, such as talc, kaolin (and other clays), chalk, zinc oxide, titanium dioxide, mica, and bismuth oxychloride. These minerals are harmless to the skin or by ingestion. They also do not cause allergies. They are only hazardous if they are inhaled from dusty products or from airbrushed makeups.

Industrial experience confirms that mineral dusts can irritate the eyes and respiratory system. Talc, kaolin, and mica can also cause long-term lung damage. In addition, talcs in the past often contained significant amounts of asbestos. Anyone who collects antique containers of baby and face powders should never use them.

No law prevents manufacturers from using asbestos-contaminated talcs. Instead, a voluntary industry standard is honored. Nevertheless, a study of commercial cosmetic talcs found traces of asbestos in six of fifteen samples.[4] But the amounts of asbestos should not be significant if people use the products without creating clouds of dust.

---

[4]Blount, A.M., "Amphibole Content of Cosmetic and Pharmaceutical Talcs," *Environmental Health Perspectives*, Vol 94, pp. 225–230, 1991.

Bismuth oxychloride added for a metallic or pearl lustre has produced photosensitivity (skin reactions provoked by sunlight) in some people. Mica can also be used for lustre.

Vegetable Powders

To avoid the hazards of mineral powders, some manufacturers have switched to organic substances, such as cornstarch or rice flour. Like all powders, these are also capable of irritating the respiratory system. They are also more likely to cause allergies, especially in people who have food allergies.

Oils, Fats, and Waxes

Cream makeups, rouges, mascaras, and lipsticks are suspended in a base made of either oils and fats alone or oils and fats emulsified with water. There are dozens of cosmetic oils, fats, and waxes. In general, natural oils, such as lanolin and cocoa butter, are more likely to provoke allergies. Almost no one is allergic to oils derived from petroleum, such as baby oil and Vaseline.

Some of the oily cream products also contain detergents, which enable the makeup to penetrate the skin for a longer hold, but which may also result in irritation.

Dyes and Pigments

Dyes and pigments are assigned names by the FDA. These names indicate whether they are approved for food, drugs, and cosmetics (e.g., FD&C Yellow #5) or only for drugs and cosmetics (e.g., D&C Red #7). Most cosmetic pigments and dyes have had long-term testing. The FDA approves of these only for uses that will not expose consumers to amounts above a threshold for causing harm.

Lipsticks contain the most dyes, and they are associated with a special form of dermatitis called cheilitis. It is a drying or cracking of the lips and usually is caused by eosin dyes, which stain the lips. Commonly used eosin dyes are D&C Red #21 and #27 and D&C Orange #5. Experts recommend avoiding these staining lipsticks. Lanolin and perfumes may also cause cheilitis.

Preservatives

Preservatives, such as thimerosal (a mercury preservative), methylparaben, and other biocides are used in small amounts, but are needed to extend shelf life and keep microorganisms from multiplying in the cosmetics. Most of these are quite toxic, but are in amounts small enough that most people will not be harmed.

ART MATERIALS

Tempera paints, watercolors, felt-tip markers, and other art materials are sometimes substituted for cosmetics for face painting or makeup effects in amateur and nonprofessional situations. However, art materials are never suitable for use as cosmetics.

The pigments, dyes, and other ingredients in art materials are not cosmetic-grade chemicals and are not safe for use on the skin. Even art materials labeled "nontoxic" are only safe when used for the label-directed artistic purpose. The person responsible for applying the art materials to the clients' skin can be held legally liable for any damages related to this off-label use.

SPECIAL EFFECTS MAKEUP

Putty, wax, beeswax, and morticians' wax all can be used to build up a part of the face for theatrical purposes. Collodion can be used to fake age or scars. Natural rubber latex can be made to function in many of these ways, and it also acts as a glue or adhesive, as does spirit gum.

"Spirit" is an old term applied to alcohol solvents (usually ethyl alcohol). Gum can mean any exudate of a number of plants or trees. Spirit gums today usually are a mixture of natural and synthetic resins in ethyl alcohol. Ideally, the resins should all be identified so users can know if they are likely to have an allergy to one or more of them.

Many people are allergic to these products. One well-known makeup artist told of a case of spirit-gum allergy severe enough to require hospital treatment. Another makeup expert avoided an actor's allergy to spirit gum by placing surgical adhesive tape on his face before applying the spirit gum. Collodion allergies also are well-known. People who are allergic to one of these products usually can find another that will do.

Spirit gum can be replaced with synthetic resin surgical adhesives in some cases. Removing spirit gum and adhesives by pulling them off the skin can be harmful, as can removing them with acetone or alcohol, which can dry or crack the skin. Some spirit gum removers are mixtures of solvents that even include toxic chlorinated solvents and skin-absorbing methanol. Take care using these products by peeling spirit gum off gently and using as little solvent as possible. Once removed, use oil, emollients, or moisturizers on the skin to counter drying effects.

NAIL PRODUCTS

Nail polishes, when used properly, are probably reasonably safe, because the amount of solvent that can be inhaled when they dry is very small. Polish removers, on the other hand, consist primarily of acetone, a solvent that can be a fire hazard and cause narcosis when significant amounts are inhaled (see chapter 11, Solvents). Although serious poisonings are rare, inhalation of acetone and other polish-remover solvents can cause headache, fatigue, and bronchial irritation.

Of greatest concern are the new liquid nail products, which harden to create long false nails. Some of these contain plastic acrylate monomers (see chapter 13 on plastics), formaldehyde, and other highly sensitizing and toxic chemicals. Take special care to apply these in very well-ventilated areas and to avoid exposure to skin and broken skin (cuts, abrasions, etc.). Should symptoms occur, discontinue use immediately.

Use of these products is also associated with nail fungus infections. These infections are extremely hard to treat and sometimes result in permanent disfigurement of the nails.

Artificial nails must never be used on stage when special effects involving fire, candles, or smoking are done on stage. The fact that long, artificial fingernails are a fire hazard was established in a study conducted at a Lamar University chemistry lab in Beaumont, Texas.[5] When in contact with a Bunsen burner flame, 87 percent of the sample nails ignited in one second or less. Even when a birthday candle was used as the ignition source, 85 percent of the nails ignited in a second or less.

All of the synthetic nails burned to completion. When victims see their nails on fire, they typically fling a hand, vigorously sending burning drops of melted plastic flying.

WHEN SKIN TROUBLE STRIKES

When skin problems arise, consult a dermatologist who can tell you which type of dermatitis you have and how to treat it. In general, if your problem is diagnosed as irritant dermatitis or cosmetic acne, you should identify the offending cosmetic and not use it again until healing is complete. If the doctor decides you are allergic to a particular product or ingredient, there are several steps you can take:

- Try hypoallergenic makeup. The term "hypoallergenic" has no legal meaning. However, reputable manufacturers honestly try to eliminate those ingredients known to produce allergies in many people.
- Try unscented makeups or products with a wholly different scent. One person in ten has an allergy to fragrances, and you may be one of them.
- Try a makeup with a different preservative. Preservatives are known to cause allergic dermatitis in some people. Three preservatives recog-

[5]Reported by *Chemical Health & Safety*, Jan/Feb 2000, p. 45 from a study by: Vanover, W.G.; Woods, J.L; Allen, S.B., *J. Chem. Educ.*, 1999, 76(11), 1521.

nized as especially hyperallergenic are Quaternium 15, imidazole idinyl urea, and parabens (both methyl and propyl parabens). Look for them on the label.

- Try comparing labels of products to which you respond, looking for an ingredient they have in common, and avoid it.

GENERAL RULES FOR MAKEUP USERS

- Use only cosmetic products on your skin. Never use paints, dyes, or other noncosmetic substances.
- Purchase only ingredient-labeled cosmetics. Many good professional theatrical brands of makeup are now ingredient-labeled.
- Use makeups only as directed. Use face makeup only on the face, eye makeup on the eyes, and so on.
- Eliminate products that contain ingredients known to cause allergies in many people, such as natural rubber, or products with toxic ingredients, such as solvents.
- Wash your hands before and after applying makeup.
- Never lend your makeup to anyone, and never borrow or accept used makeup from anyone.
- Do not use aerosol sprays or airbrush products unless there is good local ventilation in the dressing room or makeup room to remove the overspray.
- Replace old cosmetics regularly. Do not buy cosmetics that look old or shopworn.
- Avoid creating clouds of face powder or talcum, which can be inhaled. Discard old face and bath powders.
- Moisten brushes or pencils with clean tap water, not with saliva.
- Seek medical advice and treatment for eye injuries, dermatitis, acne, and other skin and eye conditions.
- Avoid smoking, eating, or drinking when applying makeup. Do not smoke or stay in dressing rooms where others smoke.
- When removing spirit gum, latex, etc., avoid prolonged skin contact with solvents like acetone. Replace lost skin oils and moisture.

ADDITIONAL PRECAUTIONS FOR MAKEUP ARTISTS

- *Ventilation.* Makeup artists are going to spend long hours in the makeup room and need to insist that the ventilation be sufficient for the products used. Most makeup rooms only have inlets in the ceiling for a recirculating ventilation system. Such rooms clearly are not ventilated sufficiently for spray and airbrush products.

  Ideally, makeup rooms should have an exhaust fan or local hood system if sprays or airbrushed makeups are going to be used. This author has seen makeup tables with slot ventilation built in. In another case, a wig room had local exhaust vents at wig-form height to catch the hair spray. These kinds of systems are rare now, but should become a standard in the industry.
- *Safe and sanitary.* Makeup artists should wash their hands before they start on each client. They should also be observant about the skin condition of their clients and use gloves if open sores, acne, or signs of skin disease are present. Gloves must be changed between clients.
- *Group makeup precautions.* Makeup artists need to ensure that their clients' makeup is not shared. Cream sticks and lipsticks can be sliced into pieces and put into small containers, labeled with each client's name. Sponges and applicators should be disposable. Powders can be supplied to each in the smallest possible containers. Eyeliners and mascara should not be shared. The water used to moisten pencils or brushes should be changed for each client. Paper cups can be used to make cleaning water containers unnecessary.
- *Training.* Makeup artists need training about bloodborne pathogens similar to the training required for home-care nurses' aides. They need to know how these diseases are transmitted, how to put on and remove gloves to avoid recontaminating themselves, how to dispose of contaminated applicators or sponges, and similar skills.

  Makeup artists also need OSHA hazard communication training about the chemicals in makeup, spray products, disinfectants, and other toxic products.

23 Teaching Theater

There is never an excuse for breaking safety and health regulations in schools. Yet, safety rules are bent and even ignored in many schools. As I see it, there are two primary reasons:

- Many teachers simply do not know the safety rules and regulations. Their administrators are not providing regular training for them as OSHA requires. As a result, they are not keeping up with the new regulations. Their curriculums now are outdated and dangerous.
- Some teachers choose ambitious productions that stress the students to their limits and use materials and effects that have requirements beyond the capacities of the shops and the venue. These teachers need to be reminded that the product of their programs is not the production; it is education. And teaching students to risk their health and break laws must not be part of any lesson.

SAFETY LAWS AND LIABILITY

Occupational health and safety laws protect teachers, because they are employees. Students (except university students in certain states) are not covered. If teachers are injured on the job, their needs are covered by workers' compensation. If students are injured or made ill by classroom activities, their usual remedy is to hold the school and/or teacher liable.

The liability of schools and teachers can best be protected by extending to students the same rights accorded adult workers. In fact, even greater care and protection is needed, since students are less trained and less educated than adult workers or teachers. This means that all occupational health or safety laws designed to protect workers doing a particular job must be complied with and exceeded if students are doing this work.

DUTIES

The types of protection that must be provided by employers for workers fall into five basic duties: to inform, to train, to enforce, to exemplify, and to provide a safe environment. The duties of teachers, then, are to meet and exceed these same duties.

- *To inform.* Just as administrators are responsible for informing teachers about the hazards of their work, teachers are responsible for passing this information on to students. The information must be complete and specific. And if a student is injured or made ill because a teacher neglected to inform her or him, the courts may interpret this as *willful* or *knowing* negligence on the part of the teacher and/or school.
- *To train.* Teachers must ensure that each student knows how to work safely with hazardous materials or equipment. Teachers must develop mechanisms to verify that students are trained, and to avoid making false assumptions about their comprehension. This is most often done by giving quizzes. Teachers should keep copies of these quizzes to document that their students understood the information and training.
- *To exemplify.* Teachers must model safe behavior. If teachers are observed violating safety rules, they may be liable for damages caused when students also break the rules. In addition, teachers must demonstrate a proper attitude toward safety. They must make it clear that safety is more important than finishing the work, cleaning up in time, or any other objective.
- *To enforce.* Teachers must be in control of their students and must enforce the safety rules. They must not tolerate improper or unsafe behavior. Teachers may be liable—even when students willfully break the safety rules—if it can be shown that the rules were not enforced. And courts have held that enforcement policies must include meaningful penalties.
- *To provide a safe shop/studio/classroom.* No amount of enforcement, training, or information will make up for teaching in an unsafe environment. *If a process cannot be demonstrated with all proper precautions, safety equipment, and ventilation, the project must be eliminated from the curriculum.* Students must never see violations of safety laws in the school.

THE ADMINISTRATOR'S ROLE

Administrators must assist and supervise teachers in performance of these duties. Administrators are ultimately responsible for the efficacy of health and safety programs, and they will share with teachers responsibility for accidents or illnesses resulting from failure to inform or train, failure to enforce the rules, or failure to set a good example.

Some administrators do not fulfill these duties under the guise of allowing teachers and students freedom. A common example is allowing university students the freedom to work alone at night in buildings without supervision, without security, and even without ventilation if it is turned off after hours. Under these conditions, administrators have no good legal defense for accidents involving chemicals or machinery, for illnesses for which emergency services were not speedily procured, or for assaults in unsecured premises.

Freedoms involving risks like these must be rescinded immediately and replaced with sound safety discipline. Administrators must administer, teachers must supervise, and students must take direction. Everyone must do their job, or the safety of all and the institution itself are put at risk.

TRAINING YOUNG STUDENTS

Students in grades six and under cannot be expected to understand the hazards of toxic substances or to carry out precautions effectively or consistently. Instead, teachers must childproof classrooms to prevent either intentional or unintentional access to toxic materials or dangerous machinery.

Youngsters are so vulnerable to toxic substances and machine hazards that university and high school students must not be allowed to bring their children into classrooms and shops. And adult theater facilities must not be used as classrooms for young children unless all adult materials and equipment are secured, all waste materials are cleaned up, and the ventilation has removed any residual toxic air contaminants.

SPECIAL STUDENTS

Access to toxic materials and potentially hazardous machinery also must be restricted for students with physical or psychological problems that impair their ability to read product labels, follow directions, or to maintain attention and self-discipline. If students with these kinds of problems are given access to hazardous materials, the liability for accidents clearly belongs to the teacher and the school.

Students who do not speak English or are illiterate also cannot be given access to toxic materials until they can demonstrate comprehension of label warnings and directions orally, in another language, or some other way. They also must not be involved in dangerous activities, such as rigging, unless they can be relied on to respond quickly to all the warning calls.

TRAINING TEACHERS

In order to properly inform and train their students, teachers need specially tailored hazard communication (HAZCOM) training (29 CFR 1910.1200). Yet, many schools still do not provide it. How ironic it is that *educational* institutions refuse to comply with a governmental mandate to *educate*.

Some administrators are slow to start training programs, because they cost money. They also fear that after they invest in training, trained faculty members will leave for other jobs. These administrators should consider the alternative: Suppose they do *not* train their teachers and they *stay?*

These are the administrators whose schools will be overtaken by increasingly complex occupational and environmental regulations and the liability they provoke. These are the schools that will graduate yet another generation of theater artists and teachers who are ignorant of the laws that apply to their work, their materials, and their health and safety.

To correct this problem, schools must take the lead. The curriculum must include formal health and safety training at levels far above the basic information required by the law. This is not only proper, it provides an opportunity for the school to develop training materials, for which there is a demand. There is good money in supplying curriculum guides, videos, and skilled trainers for business, industry, and other schools.

HAZCOM training is causing an educational revolution in the workplace. Schools should be leading the revolution, not following reluctantly.

HIGH SCHOOL AND COLLEGE STUDENT TRAINING

Many high school students already have work experience that involves HAZCOM training. Teachers often are surprised to learn that even McDonald's employees receive this training. Training is required for workers doing restaurant work, hairdressing and nails, auto body and repair work, child care, cleaning services of all kinds—any job that involves chemicals or machinery.

Students will also find that they may be more valuable to potential employers if they are already familiar with the basic HAZCOM concepts that they must learn during training. This is especially true for college students who plan to enter the profession or teach.

HAZCOM training also requires providing information about the worker's rights. Students should know that when employers do not train them, their rights have been violated. In this way, HAZCOM training in the school will protect students even off campus and in any job they take.

ADULT STUDENT TRAINING

Training of adult theater students is imperative, because teachers are bound to encounter students who know more than they do about the hazards and the laws. Sooner or later, the school will enroll a doctor or nurse, union official, plant manager, or some other professional who will know the teaching is deficient.

Adding health and safety training to classes also makes the curriculum more relevant to the students' life experience.

OTHER TRAINING LAWS

Many other OSHA regulations now require formal, documented training. Included in these regulations are:

Hazard Communication (1926.59, 1910.1200)
Respiratory Protection (1926.103, 1910.134)
Personal Protective Equipment (1926.28, 1910.132)
Emergency Plans and Fire Prevention (1910.38, 1926.150)
Fall Protection (1926.500-503)
Scaffold Regulations (1910.28 and 1926.451)
Occupational Noise Exposure (1910.95 or 1926.52)
Electrical Safety (1926.401-405, 1910.301-333)
Medical Services and First Aid (1910.151, 1926.50)
Bloodborne Pathogens Standard (1910.1030)

See chapter 2 for details about these regulations. It is clear that there is plenty of work for teachers here.

DOCUMENTATION OF TRAINING

OSHA requires documentation of training and evidence that trainees comprehend the training. Schools also want this evidence to protect their liability. The simplest method is to administer dated and signed quizzes after training sessions and keep copies in the files.

A copy of an outline of the material covered should also be on file. The following is a rather complete outline of the type of material that should be covered. This book can be a source of much of the training material.

OUTLINE OF THEATER HAZCOM TRAINING-29 CFR 1910.1200

- **Particulars of the HAZCOM law which apply to you** (see chapter 2)
  Rights and obligations, location of HAZCOM files, etc.

- **Physical characteristics of hazardous substances** (see chapter 8)
  Solids, liquids, gases, vapors, fumes, dusts, mists, and smoke

- **How materials enter the body–routes of entry** (see chapter 3)
  1. Skin contact
     Direct damage—corrosives, irritants
     Penetration through broken skin
     Absorption through normal skin
  2. Inhalation
     Absorption by lungs—gases, vapors, and soluble particulates
     Inert particulates—fibrogenic/benign pneumoconiosis
     Direct damage
     Asphyxiants—physical and chemical
  3. Ingestion
     From lung-clearing mechanisms
     Eating, smoking, drinking in the workplace
     Accidental ingestion
  4. Injection

- **Toxicological concepts** (see chapter 3)
  Dose
  Acute and chronic effects
  Total body burden
  Cumulative v. noncumulative toxins
  Multiple exposures—additive & synergistic
  Cancer, mutations, birth defects
  Sensitizers —skin/respiratory allergies

- **Factors affecting degree of hazard from toxic materials**
  Amount of material • Total body burden • Conditions of exposure • Multiple exposures • Length and frequency of exposure • High-risk groups—disabled, ill, medicated, etc. • Toxicity of materials

- **Labels** (see chapter 5)
  Use with adequate ventilation
  Danger, Warning, Caution
  Generally recognized as safe (GRAS)
  Water-based
  Biodegradable
  Low VOC
  Natural
  Hypoallergenic

Nontoxic
Under the Federal Hazardous Substances Act
Under the Labeling of Hazardous Art Materials Act

- **Detecting air contaminants**
  Sight—color, cloudiness, beams of light, etc.
  Odor—olfactory fatigue, thresholds
  Air monitoring —area sampling, personal sampling

- **Workplace Air Quality**
  TLVs and PELs (see chapter 8)

- **Outdoor Air Quality**—under EPA

- **Ventilation** (see chapter 10)
  Natural ventilation systems
  Recirculation systems—ASHRAE standards
    Air-conditioning—central and local systems
    Air-purifying systems—filters, ESPs, cyclones, etc.
  Industrial ventilation—ACGIH standards
    Dilution (aka general, mechanical) systems
    Local exhaust—hoods, ducts, fans, air cleaners

- **Substitution**
  Selecting dust-free, water-based, or low-volatility materials
  Avoiding highly toxic chemicals—carcinogens, etc.
  Comparing toxicities to find safest (LD50s, TLVs, etc.)

- **Personal protective equipment**
  Respiratory protection (see chapter 9)
  Protective clothing (see chapter 7)
    Gloves, aprons, etc.
    Face and eye protection
    Hearing protection

- **Personal hygiene and first aid** (see chapter 7)
  Eyewash fountains and emergency showers
  Hand care and washup
  Eating/recreating/living and working spaces

- **Storage and handling of materials**
  Reactive chemicals (e.g., bleach and ammonia)
  Handling of acids, alkalis, flammables
  Storage cabinets, safety cans, waste cans, etc.

- **Fire, medical, and other emergency situations** (see chapter 7)
  Extinguishers, sprinkler and deluge systems
  Emergency plans—abort procedures, drills, and training

- **Housekeeping**
  Cleanup—special vacuums, wet-mopping, etc.
  Handling spills, chemical disposal

- **Medical surveillance programs**
  Choosing doctors, medical exams, and tests

24 Safety and Americans with Disabilities

This chapter looks exclusively at health and safety issues related to the Americans with Disabilities Act (ADA). It does not address other accommodations for disabled people required by the Equal Employment Opportunity Commission (EEOC) and Architectural and Transportation Barriers Compliance Board (ATBCB) in theaters and other public buildings.

HEALTH AND SAFETY ISSUES

Some people have misinterpreted the Americans with Disabilities Act to mean that no person with a disability can be rejected for employment. This is not the case. We need to examine the actual limits on access to employment for people with disabilities.

Guidelines for employers and workers can be found in the Department of Justice's "Nondiscrimination on the Basis of Disability in State and Local Government Services" (28 CFR 35.101-190), "Equal Employment Opportunity for Individuals with Disabilities" (29 CFR 1630.1-16), the Americans with Disabilities Act "Accessibility Guidelines for Buildings and Facilities" (36 CFR 1191), and the supplementary material published with these regulations in the Federal Register (FR). These guidelines also can be applied to workers in theater, TV, film, and their employers.

UNDUE HARDSHIP

The law's Interpretive Guidance on Title 1 for Undue Hardship (29 CFR 1630.2(p)) says that: "An employer . . . is not required to provide an accommodation that will impose an undue hardship on the operation of the employer's . . . business." This undue hardship "refers to any accommodation that would be unduly costly, extensive, substantial, or disruptive, or that would fundamentally alter the nature or operation of the business."

> For example, suppose an individual with a disabling visual impairment that makes it extremely difficult to see in dim lighting applies for a position as a waiter in a nightclub and requests that the club be brightly lit as a reasonable accommodation. Although the individual may be able to perform the job in bright lighting, the nightclub will probably be able to demonstrate that that particular accommodation, though inexpensive, would impose an undue hardship if the bright lighting would destroy the ambience of the nightclub and/or make it difficult for the customers to see the stage show. (56 FR 35744-5)

This example is relevant to work in theater, where people with similar visual impairments can be excluded from tasks that require them to maneuver backstage during blackouts or seat people after the curtain is up. However, any disability that would interfere substantially with the very nature and financial viability of a scenic artwork will qualify. Examples might include color blindness and an inability to climb ladders.

DIRECT THREAT

In the Interpretive Guidance on Title 1 for Direct Threat (29 CFR 1630.2(r)) it says that: "An employer may require, as a qualification standard, that an individual not pose a direct threat to the health or safety of himself/herself or others." This text should be read in its entirety to understand when the threat is considered significant. However, I call attention to the following:

> For example, an employer would not be required to hire an individual, disabled by narcolepsy, who frequently and unexpectedly loses consciousness, for a carpentry job the essential functions of which require the use of power saws and other dangerous equipment, where no accommodation exists that will reduce or eliminate the risk. (56 FR 35745).

Other commonly accepted restrictions on jobs might include hearing impaired people working on rigging (a job in which hearing warning calls without visual contact is important) or people with epilepsy working on any job involving operation of dangerous machinery or motor vehicles.

Under the rule, it is not easy to find out when a risk to a disabled worker or other workers is substantial:

> The assessment that there exists a high probability of substantial harm to the individual, like the assessment that there exists a high probability of substantial harm to others, must be strictly based on valid medical

> analyses and/or on other objective evidence. This determination must be based on individualized factual data. . . . (56 FR 35745)

The problem here is that most employers do not require the disclosure of medical conditions or limitations on which such decisions must be based. It is in this gray area that many disputes arise over when a disability is substantial.

For this reason, qualification standards and descriptions of the physical requirements for various jobs should be developed. And these descriptions must be detailed. For example, a position only described as "Scenic Artist" would not be informative. Instead, the *physical* description of the work might include:

> SCENIC ARTIST: Must be able to discern subtle differences of color, climb ladders and work safely on scaffolds, lift paint containers and equipment up to a weight of fifty pounds, tolerate exposure to chemicals in products used in the scene shop at levels below the OSHA-permissible exposure limits, wear respiratory protection when needed, etc. . . .

The description can be accompanied with either a disclosure requirement or a disclaimer facilitating firing of individuals who knowingly take jobs for which they do not meet the physical qualifications. To enable employers to develop job descriptions, Appendix A of the law includes an "Essential Functions Information Form." I suggest getting a copy of this form and filling it out while observing someone already doing a job.

Other job/class qualifications can take the form of licenses or certifications. Examples of "safety qualifications that would be justifiable in appropriate circumstances" in the "Section-by-Section Analysis" of 28 CFR 35.130 include:

> . . . eligibility requirements for drivers' licenses, or a requirement that all participants in a recreations-rafting expedition be able to meet a necessary level of swimming proficiency. (56 FR 35705)

PEOPLE WITH ALLERGIES

In a discussion of the definition of "substantially limits" in 29 CFR 1630.1(j), the Equal Employment Opportunity Commission identifies "allergy" as a disability:

> . . . Suppose an individual has an allergy to a substance found in most high-rise office buildings, but seldom found elsewhere, that makes breathing extremely difficult. . . . [This] individual would be substantially limited. . . . (56 FR 35742)

Since serious allergies are a recognized disability, job descriptions should include the types of chemical products that will be used. Otherwise, for example, workers allergic to solvents and paint chemicals conceivably could demand that all use of these products cease where they work! Further, the employer would have to accommodate them with alternative materials or alternate work.

Workers who first develop an allergy to a chemical on a particular job cannot be fired because they can no longer perform the task for which they were hired. However, in our business, jobs usually are of short duration. This makes proving the source of an allergy almost impossible. Worse, once the job is over, employers are not under any obligation to hire the now-disabled worker for any future production.

This is especially a problem for performers and orchestra musicians who develop allergies to special-effects chemicals. If employers notify all new applicants that special effects will be used, they technically do not have to accommodate workers who are already allergic to them.

The problem of workers who are allergic to special effects must be addressed jointly by unions and employers in a humane and reasonable manner. These workers now are very limited in the types of jobs they can take. Something needs to be done for them by the industry that disabled them.

REPRODUCTIVE HAZARDS

Job descriptions that include potential chemical exposures may also be used to alert workers planning a family. While there are no laws to specifically protect the developing fetus, employers may be sued on behalf of a child born with defects or health problems caused by chemicals used on the job. Clearly, these suits are a poor substitute for a healthy child.

Solvents used in paints and other products have now been associated with birth defects in one study. While more research is needed to quantify this hazard, couples planning a family cannot wait for the data. There is no good answer for this problem other than turning down jobs and trading a chemical problem for an economic one.

MULTIPLE CHEMICAL SENSITIVITIES (MCS)

While allergy is discussed by the EEOC, MCS is specifically excluded from consideration by the ADA. In the General Issues section of the supplementary information provided with 36 CFR.1991, it says that during the proposed rule's comment period, the Architectural and Transportation Barriers Compliance Board received over four hundred comments from individuals who identified themselves as chemically sensitive.

The commenters described health problems from indoor contaminants, such as those from building materials, furnishings, cleaning products, and fragrances. They suggested lessening exposure by providing windows that open, improving heating and ventilation, and selecting building materials and furnishings that do not contain certain substances. The board's response was:

> Chemical and environmental sensitivities present some complex issues which require coordination and cooperation with other Federal agencies and private standard setting agencies. Pending further study of these issues, the Board does not believe it is appropriate to address them at this time. (56 FR 35412)

This seems to indicate that accommodation is not required for individuals whose claims of disability from exposure to chemicals cannot be supported by traditional allergy testing or other accepted medical-diagnostic procedures. There are some court actions in which individuals claiming MCS have won the right to be accommodated. However, such accommodations are rare.[1]

OTHER APPLICABLE LAWS

OSHA regulations also require certain physical abilities on the part of workers doing hazardous work. For example, jobs for which respirators must be worn require that workers are healthy enough to wear a respirator (29 CFR 1910.134). This is usually determined by a doctor's written authorization or a detailed medical questionnaire. (Workers required to wear respirators also must come to work clean-shaven, or they can be legally dismissed.)

EMERGENCY EGRESS

In most cases, disabled individuals who meet job description requirements cannot be excluded because of architectural barriers, such as stairs. For this reason, disabled people's needs must be considered when building or renovating shops and studios.

Access is only half of the problem. Egress is the other half. There is no point in getting disabled people into a facility if they cannot be easily and quickly evacuated in an emergency. Unfortunately, this fact was overlooked in the rush to provide access. Only recently has the National Fire Protection Association begun revising their "Life Safety" standard (NFPA 101) to provide guidelines for egress of disabled people. All construction and renovation plans must be assessed for egress of disabled people.

---

[1]Multiple Chemical Sensitivity as a "handicap" requiring accommodation is more accepted on a case-by-case basis by the U.S. Department of Housing and Urban Development's Fair Housing Act, but not in employment situations.

Existing facilities also must be assessed to assure that disabled people are not allowed into areas from which they cannot be rapidly evacuated in an emergency. For example, people whose life support system is in their wheelchair (which means they cannot be lifted from their chairs and carried to safety) must not be allowed on floors with heat-seeking elevators.

SUMMARY

In this chapter, I concentrated on the negative aspects of accommodation. On the positive side, I have observed that opening access to many types of jobs has been positive overall. Some of the benefits include reducing the physical tasks, hazardous chemicals, and dangerous equipment presently in use. The substitution of safer materials and projects results in providing environments that are healthier and safer for all workers.

Scene painting and construction, on the other hand, does entail certain physical requirements that cannot be met by some types of disabled people. For emotional and humanitarian reasons, we may want to open jobs to everyone. But in this case, we would endanger the very people we are trying to help.

Bibliography and Sources

ARTS, CRAFTS, AND THEATER SAFETY (ACTS)

ACTS is a nonprofit corporation dedicated to providing health and safety services to the arts. Unlike many other not-for-profits, we do not actively solicit donations from the people we serve. We recognize that artists and performers are among the least affluent groups in society.

ACTS also does not accept donations or advertising from businesses or manufacturers. Despite these restrictive policies, we generate sufficient income from unsolicited contributions from individuals and foundations and from below–market-value fees for services including:

- providing speakers for lectures, workshops, and courses;
- conducting U.S. and Canadian OSHA/OHSA compliance training sessions and inspections;
- providing technical assistance for building planning, renovation, and ventilation projects;
- researching, writing, and editing safety materials; and
- sale of publications (see list in the Bibliography).

This income is used to support our below-cost and free services:

- A worldwide free information service by phone, mail, and e-mail providing: professional safety and industrial hygiene advice; copies of educational materials; access to research materials from our extensive technical library; referrals to doctors and other sources of help; and

- the *ACTS FACTS* newsletter which is provided at the cost of first-class postage and reproduction.

We maximize the use of this income by keeping our overhead low. We operate from our own home offices, communicate by e-mail and conference calls, and laser print/copy our own publications. Questions, comments, and requests for our services are welcome.

SAFETY SERVICES FOR UNION MEMBERS

As the safety director for the United Scenic Artists, Local 829, International Alliance of Theatrical and Stage Employees, Monona Rossol is authorized to provide publications and certain kinds of training to theatrical union members at no cost. Call, write, or e-mail ACTS, but identify yourself and your union.

ACTS DATA SHEETS

Data sheets are available from ACTS. Call, write, or e-mail ACTS for terms.

A Hazard Communication Program*
    a fill-in-the-blanks program for employers
All About Wax
Americans with Disabilities in the Scenic Arts*
Anthraquinone Dyes & Pigments
Artist's Oil Painting
Assessing Locations Hazards*: 3 similar versions for
    Scenic Artists
    Scouts and Managers
    Actors and Cameramen
Azo Dyes & Benzidine Dyes, banned German dyes
Biological Hazards

Carbon Arcs: Still Around
Carbon Monoxide (CO) Detectors*
Carpets: The Case Against Them
D-Limonene: the Citrus Solvent
Dyes and Pigments
Gloves—Training Materials*
    types of gloves, OSHA rule, & rubber hazards
Heat Exposure in shops in the summer*
Identifying Dyes
Identifying Pigments
Labels: Reading Between the Lies
Lead-containing Art & Theater Materials
Lead: Effects on Adults & Children
Lead Laws and the Scenic Arts*
Linseed Oil: A Fire Hazard
Mold in Costume Storage Areas
Mold: Nothing to Sneeze At
MSDSs: Reading Between the Lines
OSHA Regulations and the Scenic Arts*
OSHA Regulations that Apply to Costume Work*
Oven-Cured Polymer Clays
Reproductive Hazards for Scenic Artists
Respiratory Protection: New Rules*
Respiratory Protection Program*
    a fill-in-the-blanks program for employers
Selecting Children's Art Materials
Solvents*
Teaching Art & Theater Safely
Threshold Limit Values: a simplified definition*
Urethane Resin Systems*
Using Artists Paints for scenic work*
Understanding the MSDS
Ventilation for Theaters and shops*

*available courtesy of United Scenic Artists

GOVERNMENTAL AGENCIES

Occupational Safety training and compliance help

Occupational Safety and Health Administration (OSHA)
General Industry Compliance Assistance
Directorate of Compliance Programs
200 Constitution Avenue, N.W., Room N-3107
Washington, DC 20210
(202) 693-1850
www.osha.gov

Environmental and waste disposal regulations

Environmental Protection Agency (EPA)
www.epa.gov

Product safety regulations

U.S. Consumer Product Safety Commission (CPSC)
Washington, DC 20207
(800) 638-2772
www.cpsc.gov

Especially useful for people purchasing costumes, props, and equipment is the list of recalled products in their "Thrift Store Checklist."

Occupational safety and health standards and publications

National Institute for Occupational Safety and Health (NIOSH)
4676 Columbia Parkway
Cincinnati, OH 45226-1998
(800) 35-NIOSH
E-mail:pubstaft@cdc.gov
www.cdc.gov/niosh

New Jersey Department of Health—Right to Know Program

New Jersey Department of Health's Right to Know Hazardous Substances Fact Sheets provide better information than you will find on MSDSs including more detailed chronic data, whether the chemical has ever been formally studied for cancer effects, and odor thresholds when known. They are available for $1.00 each in hard copy. The whole database also is available on CD-ROM and online services. Contact:

New Jersey Department of Health and Senior Services
Right to Know Program—CN 368
Trenton, NJ 08625-0368
(609) 984-2202

Or download individual data sheets at:
www.state.nj.us/health/eoh/rtkweb/rtkhsfs.htm

STANDARDS ORGANIZATIONS

Industrial air quality and industrial ventilation standards

American Conference of Governmental Industrial Hygienists (ACGIH)
1330 Kemper Meadow Drive
Cincinnati, OH 45240-1634
Publications catalog: (513) 661-7881
www.acgih.org

Especially useful is their *Industrial Ventilation: A Manual of Recommended Practice* and their yearly updated booklet of Threshold Limit Values.

Comfort ventilation and some indoor air quality standards

American Society of Heating, Refrigerating and Air-conditioning Engineers (ASHRAE)
1791 Tullie Circle, N.E.
Atlanta, GA 30329-2305
(404) 636-8400
www.ashrae.org

Fire safety, life code, pyrotechnic, and other standards

National Fire Protection Association (NFPA)
1 Batterymarch Park
PO Box 9101
Quincy, MA 02269-9101
(617) 770-3000
www.nfpa.org

Some useful standards include:

NFPA #70 National Electrical Code
NFPA #102 Assembly Seating, Tents, Air-Supported Structures
NFPA #1123 Fireworks, Public Display
NFPA #1126 Pyrotechnics Before a Proximate Audience
NFPA #160 Flame Effects Before an Audience

Product safety standards including the standards for art materials, ASTM D-4236

American Society of Testing and Materials (ASTM)
100 Barr Harbor Drive
W. Conshohocken, PA 19428-2959
Publications (610) 832-9585
www.astm.org

Welding standards including safety standards and training courses

American Welding Society (AWS)
550 N.W. LeJeune Road
Miami, FL 33126
(800) 443-9353
www.aws.org

Industry standards for special effects

Entertainment Services & Technology Association (ESTA)
875 Sixth Avenue, Suite 2303
New York, NY 10001
(212) 244-1505
www.esta.org

Safety standards for the film industry

American Alliance of Motion Picture and Television Producers (AMPTP)
Industry Wide Labor-Management Safety Committee
15503 Ventura Boulevard
Encino, CA 91436-3140
(818) 995-0900

Especially useful are their "33 Safety Bulletins" which provide brief rules for use of firearms, fog and smoke, helicopters, lighting systems, and many other hazardous materials and equipment.

COMMERCIAL SOURCES

The companies listed here are not the only ones providing services and inclusion does not constitute an endorsement of these companies or their products.

SAFETY SUPPLIES: Unless you are an expert, use a safety catalog supplier that provides a wide array of products and technical help in making your selections. Two such companies are:

Lab Safety Supply
Technical Support: 800/356-2501
order: (800) 356-0783
info: *www.labsafety.com*

Direct Safety
(800) 528-7405
E-mail: directsaf@aol.com

TRAINING VIDEOS AND INSTRUCTIONAL MATERIALS

For training materials especially for theater

Theater Arts Video Library
174 Andrew Avenue
Leucadia, CA 92024
(800) 456-8285
www.theatreartsvideo.com

For general training materials

Long Island Productions
106 Capitola Drive
Durham, NC 27713
Phone: (800) 397-5215
Fax: (919) 544-5800
E-mail: longislandproductions.com

Coastal Safety & Environmental
3083 Brickhouse Court
Virginia Beach, VA 23452
(800) 767-7703
www.coastal.com

CONSULTANTS

Industrial hygienists and certified testing laboratories

American Industrial Hygiene Association (AIHA)
2700 Prosperity Avenue, Suite 250
Forfeits, VA 22031
(703) 849-8888
www.aiha.org

Rigging inspectors and trainers

Sapsis Rigging
233 North Lansdowne Avenue
Lansdowne, PA 19050
(800) 727-7471
www.riggingseminars.com

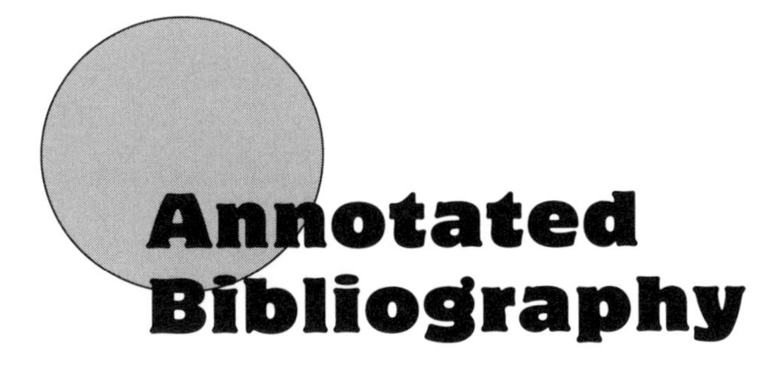

Annotated Bibliography

Clark, Nancy, Thomas Cutter, and Jean-Ann McGrane. *Ventilation: A Practical Guide.* Center for Safety in the Arts, New York, 1980. A guide to basic ventilation principles and step-by-step guidance for those who wish to evaluate, design, and build an adequate ventilation system. Available on Amazon.com.

Colour Index International, Society of Dyers and Colourists (P.O. Box 244 Perkin House, 82 Grattan Road, Bradford West Yorkshire BD1 2JB England, and the American Association of Textile Chemists and Colorists (One Davis Drive, P.O. Box 12215 Research Triangle Park, NC 27709-2215, U.S.A.), Volumes 1-4 and 6-9 (Cost about $900). Lists by C.I. name and number all commercially available dyes and pigments and their intermediates. Every degree program that involves use of paints, dyes, pigments, or colorants should include instruction on how to use this reference and the C.I. system of names and numbers.

Dryden, Deborah M. *Fabric Painting and Dyeing for the Theatre.* Portsmouth, NH: Heinemann, 1993. Well-written, 256-page, practical guide for dyers with accurate safety and chemical information.

Haas, Ken. *The Location Photographer's Handbook: The Complete Guide for the Out-of-Studio Shoot.* New York: Van Nostrand Reinhold, 1989.

Hawley, Gessner. *Hawley's Condensed Chemical Dictionary,* 13th Ed., revised by N. Irving Sax and Richard Lewis, Sr. New York: Van Nostrand Reinhold,

1992. (Also available from the ACGIH. Call (513) 661-7881 for publications catalog.)

Patty, Frank, ed. *Industrial Hygiene and Toxicology,* Vols 1-9, 5th ed., New Jersey: John Wiley & Sons, Inc., 2000. Very expensive ($250/volume) and very technical, but a classic in the field.

Rossol, Monona. *The Artist's Complete Health and Safety Guide,* 2nd ed., New York: Allworth Press, 1994. Winner: Choice Outstanding Academic Book Aware, Assoc. of College and Research Libraries. A guide to safety and OSHA compliance for those using paints, pigments, dyes, metals, solvents, and other art and craft materials. Available from ACTS, or for credit card orders: 1-800-491-2808.

Sax, N. Irving. *Dangerous Properties of Industrial Materials,* 8th ed. New York: Van Nostrand Reinhold, 1993. A new addition of this is just out. The cost is in the range of $1000. A good fast reference, but only a starting point.

FOG/SMOKE/PYROTECHNIC PUBLICATIONS

WRITTEN BY MONONA ROSSOL

STAGE FRIGHT: Health and Safety in the Theater. New York: The Center for Occupational Hazards, 1986. (Reprinted, ACTS, 1987. Republished, Allworth Press, 1991.) Probably the first book written exclusively on theatrical occupational safety and health.

Danger, Artist at Work. Australia Council, 1990, Second Edition, 1995. Theatrical hazards included.

"Ill Effects of Theatrical Special Effects." *Medical Problems of Performing Artists,* Vol. 1, No. 2, June 1986. This was the first medical journal article on this subject.

"Musician's Hazards." *Senza Sordina* (Publication of the Conference of Symphony & Opera Musicians), Vol. 28, No. 4, April 1990.

"Theatrical Fogs and Smokes: A Report on Their Hazards." New York: American Guild of Musical Artists, in cooperation with AEA, SAG and AFTRA, 1991. The national secretary of SAG says that this report is now used as a reference in every country in the world where film or television commercials are made.

"Health Effects from Theatrical Pyrotechnics." *The Journal of Pyrotechnics.* Kosanke Publishers, Issue No. 3, Summer 1996.

"Artificial Fogs and Smokes," co-authored with Harry Herman, and "Pyrotechnics," chapters for 2nd Ed. of *Professional Voice: The Science and Art of Clinical Care,* ed. by Robert Thayer Sataloff, MD. San Diego: Singular Publishing Group, 1997.

"Another Fog Story." *Journal of Pyrotechnics,* Issue 7, Summer 1998.

"Review of 'Pollution Caused by Fireworks.'" *Journal of Pyrotechnics,* Issue 8, Winter 1998, pp. 74–75.

"Review of book *'Theater of Fire,'*" P. Butterworth, *Journal of Pyrotechnics,* Issue 9, Summer 1999.

Also on request, dozens of articles, first in the *Art Hazards News* (first one in 1983) and in *ACTS FACTS,* from 1987 to present.

OTHER AUTHORS

NIOSH Health Hazard Evaluation Technical Assistance report (HETA 82-365-1282), Las Vegas, NV: MGM Grand Hotel & Casino, 1982. Air and dust sampling of pyrotechnic smoke.

NIOSH Health Hazard Evaluation Technical Assistance report (HETA 84-478), 1984. Air sampling of inorganic chlorides from "smoke cookies."

Alliance of Motion Picture and Television Producers. *SAFETY BULLETIN NO. 10: Guidelines Regarding the Use of Artificially Created Smokes, Fogs and Lighting Effects*. Revised June 1992.

NIOSH study began in 1990 on complaints of health effects from fog by Actor's Equity. The report went through four versions: a draft, an interim report, a revised interim report (HETA 90-355), and a final report released in August 1994 (HETA 90-355-2449). The final report appends the revised interim report.

A fifteen-year-old girl with mild asthma experienced an exacerbation of her asthma after being exposed to artificial fog during a theater rehearsal. *The Lancet,* January 7, 1994.

Herman, Jr., Harry H. "Health effects of glycol based fog used in theatrical productions." Report to Actor's Equity Association, July 1995. (presented in April 1995 at the American Chemical Society's annual meeting)

Herman, Jr., Harry H. "Are Theatrical Fogs Dangerous?" *Chemical Health & Safety*. American Chemical Society, (2)4, July/August 1995, pp. 10–14.

Evans, Randolph W. MD, Richard I. Evans, PhD, Scott Carvajal, MA, & Susan Perry, PhD. "A Survey of Injuries among Broadway Performers." *Am. Journal of Public Health,* 86(1), Jan 1996, pp.77–80. Not about fog exclusively, but notes that 17% of the injuries to actors are vocal and fog is commonly cited as the cause.

"Introduction to Modern Atmospheric Effects." ESTA (Entertainment Services & Technology Association), 1996. For copies call Broadway Press, 800-869-6372 and send $7.95.

Neumann, Heinz Dieter, Jens Uwe Hahn, Heinz Assenmacher—Maiworm, Heinrich Birtel, and Kussin Heike. Chemical Abstract #126:108062f Exposure to fog fluids used on theater stages. (GUVV Wesstfalen-Lippe, D-48159 Muenster, Germany). *Gefaharstoffe-Reinhalt. Luft* 1996, 56(11), 431-436 (Ger), Springer. Air and material samples were analyzed as a result of numerous complaints about irritative symptoms after contact with fog fluids put forward by employees of a theater. The irritations were suspected to be due to aldehydes occurring during pyrolysis of fog fluids in fog generators.

Report on a study of about 25 Local 802 pit musicians at *Beauty and the Beast* on Broadway by Dr. Jacqueline M. Moline, Mount Sinai-Irving J. Selikoff Center for Environmental and Occupational Medicine, January 17, 1997.

Ruling, Karl G. "Fog Safety Still a Concern for Users and Manufacturers." Entertainment Design, July 2000, pp. 20–23. Mr. Ruling is the technical manager for ESTA and states their position.

Index

BOOKS FROM ALLWORTH PRESS

Technical Theater for Nontechnical People by Drew Campbell (softcover, 6 × 9, 256 pages, $18.95)

Clues to Acting Shakespeare by Wesley Van Tassel (softcover, 6 × 9, 208 pages, $16.95)

Casting Director's Secrets: Inside Tips for Successful Auditions by Ginger Howard (softcover, 6 × 9, 208 pages, $16.95)

Creating Your Own Monologue by Glenn Alterman (softcover, 6 × 9, 192 pages, $14.95)

Promoting Your Acting Career by Glenn Alterman (softcover, 6 × 9, 224 pages, $18.95)

An Actor's Guide—Your First Year in Hollywood, Revised Edition by Michael Saint Nicholas (softcover, 6 × 9, 272 pages, $18.95)

Producing for Hollywood: A Guide for Independent Producers by Paul Mason and Don Gold (softcover, 6 × 9, 272 pages, $19.95)

Writing Scripts Hollywood Will Love, Revised Edition by Katherine Atwell Herbert (softcover, 6 × 9, 160 pages, $14.95)

Making Independent Films: Advice from the Filmmakers by Liz Stubbs and Richard Rodriguez (softcover, 6 × 9, 224 pages, $16.95)

Get the Picture? The Movie Lover's Guide to Watching Films by Jim Piper (softcover, 6 × 9, 256 pages, $18.95)

The Screenwriter's Guide to Agents and Managers by John Scott Lewinski (softcover, 6 × 9, 256 pages, $18.95)

So You Want to Be a Screenwriter: How to Face the Fears and Take the Risks by Sara Caldwell and Marie-Eve Kielson (softcover, 6 × 9, 224 pages, $14.95)

Please write to request our free catalog. To order by credit card, call 1-800-491-2808 or send a check or money order to Allworth Press, 10 East 23rd Street, Suite 510, New York, NY 10010. Include $5 for shipping and handling for the first book ordered and $1 for each additional book. Ten dollars plus $1 for each additional book if ordering from Canada. New York State residents must add sales tax.

To see our complete catalog on the World Wide Web, or to order online, you can find us at *www.allworth.com*.